Continuous Delivery Blueprint

How to implement efficient software change management
processes at the enterprise level in the era of clouds,
microservices, automation, and DevOps.

Max Martynov and Kirill Evstigneev

Continuous Delivery Blueprint

by Max Martynov and Kirill Evstigneev

ISBN-13: 978-0-9600271-0-1

This book would not have been possible without the support and help of many people. We are very grateful to our colleagues and friends, Max Shishkarev, Stan Klimoff, Victoria Livschitz, Sunil Bhardwaj, Jimit Ladha, Sergey Plastinkin, and Dmytro Zamaruiev, who helped forging the ideas presented in the book during brainstorming sessions and implementation projects, reviewed the content of the book, and offered their feedback. Special thanks go to Anastasia Martynova, who reviewed the first early draft of the book and gave extremely valuable feedback.

We are indebted to Victoria Livschitz, Leonard Livschitz, and Ezra Berger for supporting this project and helping with the publishing. Last, but certainly not least, thanks to Kathryn Wright, the editor, and Mykhaylo Bolotov, the designer, who helped us shape the book into the final product.

"Max's book came in right on time, as we are in the midst of developing our Managed Service Offering here at GridGain Systems. As I was reviewing the draft, it became clear that this is a far more canonical and in-depth look at modern DevOps and development processes than most other books or online resources can provide. The book covers a wide swath of material, but in the end, it manages to deliver a comprehensive and well thought-out picture of modern software development processes."

— Nikita Ivanov, Founder & CTO, GridGain Systems

"This book provides a definitive guide on how to build Continuous Delivery practices that automates the processes between software development and IT teams, enabling them to build, test, and release software faster and more reliably. The principles laid out in this book provide us with a roadmap to accomplish these goals."

— Lance Wills, eCommerce EVP

"This book by Max Martynov and Kirill Evstigneev draws from deep domain expertise they developed at Grid Dynamics in building effective DevOps teams and delivering Continuous Delivery pipelines for large organizations to help them successfully, accelerate their application development projects, improve quality and bring digital services to the market faster."

— Eric Benhamou, Founder and General Partner,
Benhamou Global Ventures

"Technology advancements in cloud infrastructure, containerization, architecture, and automation are finally making it possible to implement a reliable Continuous Delivery process and build DevOps best practices. This book starts with an overview and finishes with a detailed recipe on how to utilize modern technologies and processes to build efficient and reliable change management in a company. Even if your organization is already in the process of implementing Continuous Delivery, the book may provide more interesting and practical ideas on how to improve upon what you have now."

— Kira Makagon, Chief Innovation Officer, RingCentral

CONTENTS

ABOUT THE BOOK

The motivation to write this book came from an observation that the existing literature on DevOps and continuous delivery either describes how to implement it at a small scale, a single application or a single team, or focuses primarily on the cultural aspects of DevOps, or describes a success story that may be specific to a company and an environment.

This book attempts to bridge the gap by establishing an organizational and technical framework for the implementation of DevOps and continuous delivery in large companies. The framework is created to satisfy the following requirements:

- Scalability: support IT organizations with thousands of members and systems consisting of hundreds of services.

- Efficiency: support multiple concurrent changes to production, with the deployment of tens of changes to production per service during a day.

- Quality, controls, and auditability: support high quality and all of the necessary controls and audit logs that a typical enterprise requires.

- Transparency and KPIs: measure how much time and effort each step in the process takes to be able to optimize the process going forward and to recognize offending components and teams.

The book is structured to show how to transform the change management process in an existing large IT organization to adopt clouds, microservices architecture, DevOps, and continuous delivery. Each chapter describes a single topic or idea. For each topic, we first present a conceptual overview of the problem and a recommended solution. We then describe an example of implementation, as well as the options for technologies and tools that can be used to implement a solution. In some cases, we believe that there is no good tool on the market that can solve the problem, and in this case, we outline the features that need to be implemented. These features can be taken for custom implementation. In the future, we hope that the necessary tools will be created in the industry and all gaps will be closed. For visual readers, we've included graphics, diagrams, and charts to better explain complex concepts and processes.

The book was created as the result of multiple engagements in building robust and efficient continuous delivery processes in start-up and enterprise environments at different scales: from dozens of engineers to thousands of engineers, and from several applications to hundreds of applications. Many recommendations in the book represent the opinions of the authors based on experience. We recognize that there may be other approaches to solve a problem. In some cases, we explicitly mention other approaches with their pros and cons, and in other cases, we omit a comprehensive review to keep the volume of the book reasonable. Sometimes an approach, which we do not recommend in general, may work well in very specific circumstances. If, after analyzing our recommendations, you recognize that a different approach would work better for you or you already have a non-recommended approach working well in your organization, keep using what works. It is important to understand that the task of establishing a modern continuous delivery process consists of many small solutions that can be implemented independently. If you do not agree with one or several recommendations presented in the book, you may still find other recommendations useful. The authors would appreciate it if you would share your experience, especially if it is different from the approaches recommended in the book.

TARGET AUDIENCE

The book will be helpful to:

- Executives in enterprises and large organizations, who are tasked with migration to the cloud and microservices architectures with open source technology stacks, while reducing the time-to-market and enabling continuous delivery and DevOps best practices. The book will give them a framework to transform their organizations and processes to accommodate the migration.

- Directors and managers in traditional IT organizations performing service delivery functions like release engineering, production support, and non-production environment maintenance. The book will describe how to redesign processes to implement existing policies in a continuous delivery fashion.

- Developers, QA engineers, and build and release engineers who are involved in the software development lifecycle. The book will explain how to extend CICD pipelines all the way to production, but at the same time satisfy the requirements of the traditional release engineering and production operations teams.

In essence, the book attempts to bridge the gap between development and operations and give a framework that can work at the enterprise scale. After reading the book, developers will be familiar with the operations team's way of thinking, the operations team will be able to create a new and modern framework to facilitate rapid development, and managers will have a better idea of how to reorganize their teams to support transformation.

1

INTRODUCTION

1.1. TRANSFORMATION DRIVERS

There are several major business drivers for enterprises and large organizations with traditional IT teams to transform the way they undertake change management and production operations:

1. Decrease the time-to-market and enable faster and more frequent release cycles. Customers want new features quickly, and whoever delivers features first wins. So, just as traditional consumers want their goods to be delivered the same day, business is pushing technology and development teams to deliver features as quickly as possible. Development, in turn, pushes operations to release features faster and more frequently than ever before. A shorter time-to-market also helps to improve the return on investment from development activities by shipping features faster and getting value out of those features sooner.

2. Decrease and optimize operational costs. A traditional change management process in a big IT organization involves many people from different departments to review and approve a particular change. Things are worse when such reviews need to happen for multiple concurrent changes: this may require several forms to be submitted, several meetings to be organized, new teams to be created (a change control board), ongoing calls to be made, and follow-up emails to be sent in order to move things forward. IT operations departments are under constant pressure to reduce the cost of change management and production operations without a decrease in quality and control.

From the technology perspective, there are several technological and cultural innovations that are pushing traditional IT teams to change:

1. The DevOps movement. DevOps promises higher efficiency and a faster time-to-market by introducing a close collaboration between the development and operations roles in an attempt to destroy the "wall of confusion" [1], so IT teams are being pushed to accept this methodology. In some cases, this push comes from development teams that don't necessarily understand all of the risks and intricacies of production operations, which leads to conflict instead of collaboration. Often, the push comes from top management who are tired of constant fights between two departments with different objectives: development needs to make a change, whereas operations needs to provide stability. Besides pure efficiency gains, the implementation of DevOps best practices gives engineers end-to-end responsibility for the final result of delivering customer-facing features to production. This, in turn, leads to higher motivation, job satisfaction, and productivity [2].

2. Public clouds. Many enterprises are undergoing massive migration to public clouds in order to get rid of data-center management requirements and reduce operating expenses. Public clouds bring way more flexibility than classical data centers but require a completely different mentality for dealing with infrastructure. Whereas virtual machines (VMs) were previously managed as "pets," they now need to be treated as "cattle" [3].

3. Open source technologies and microservices architectures. Big companies are replatforming from expensive monolithic out-of-the-box products provided by Oracle, IBM, SAP, and Microsoft to custom platforms built with open source technologies. The adoption of microservices, or at least well-designed service-oriented architectures with reasonable granularity of services, leads to the management of hundreds or thousands of services instead of tens.

4. Automation. The heavy automation of testing and adoption of container technology for build and packaging, application deployment, infrastructure provisioning, and process orchestration outperforms old manual processes and procedures in efficiency.

With all of the drivers above, IT organizations that are using traditional IT Service Management (ITSM), IT Infrastructure Library (ITIL), and Control Objectives for Information and Related Technologies (COBIT) processes are under massive pressure to become more efficient without loss of quality and controls. In a way, just as IT organizations are managing change when it comes to software, they now need to change themselves according to the new requirements from business and the new technology enablers from the industry.

1.2. WHY IS IT DIFFICULT?

Why are there many success stories in startup environments and not that many in enterprise environments, even though some of the IT organizations in those startups are now getting close in size to the enterprise ones?

The reason is that the startups began their journey without the legacy of old processes. When they started rolling out the first features in production, their risks related to instability or delivery of defects were extremely small. They just needed to push more features to the customers. That is, the speed for the delivery of new features outweighed the risks by a large margin. Their production support teams were tiny, and the developers were very close to production from the beginning.

Of course, once those startups reach a large scale, they need to start dealing with all the same problems that the traditional enterprises have. But by that time, their development teams already have several years of experience of being close to production and partially responsible for quality and production operations. They already have organically established procedures that are efficient and produce reasonably stable results. When they need to start establishing stricter controls and passing external audits, they just adapt their existing processes to the controls for those audits. Essentially, the startups' IT organizations end up in a state where they have all the controls and policies that traditional enterprises have, but those controls and policies have been implemented with much more efficient procedures and processes and use modern technologies. The path to efficiency is not easy however: they will have undergone several years of trial and error with instabilities and defects in production until their processes have evolved.

This approach wouldn't work for enterprises. Big companies have an established business with a low tolerance to risks, as well as an already developed organizational structure with long-standing processes, procedures, and culture. Those processes, procedures, and culture have been forged over tens of years of working in a different environment with alternative objectives, technologies, and constraints.

The traditional model of responsibilities for different stages of an application lifecycle that most enterprises operate is shown in Figure 1.1. In a nutshell, the development of custom code is shadowed by management of out-of-the-box products, and the development teams are only responsible for writing code. We will explore the origins of this model in the Starting Point chapter later in the book. All of the change management steps after the code is written, including build, continuous integration, testing, non-production environment management, infrastructure provisioning, application deployment, release engineering, production deployment, and production operations, are done by an IT operations organization. This organization may have different names, and it may have multiple sub-organizations, such as infrastructure and quality assurance (QA). The key is that it is separate from development. Traditionally, most of the functions in the operations domain are not well-automated, including testing, deployment, provisioning, etc.

Figure 1.1. Traditional model of responsibilities at different stages in an application lifecycle.

To deal with the pressure for speed and efficiency, enterprises are trying to implement Agile and DevOps methods. This is typically done in one of two ways:

1. Select a standalone project and implement DevOps for it.

2. Expand development's area of responsibility by letting them implement and automate the continuous integration and continuous delivery (CICD; see the Definitions chapter) pipeline, tests, and deployments.

Unfortunately, our experience shows that both ways have their own challenges and tend to fail if not done with extreme care and attention to detail. When a standalone project is created, DevOps best practices and continuous delivery may actually be implemented in it and a new shadow IT may be created there. But when the project grows to a certain size, it appears on the operations team's radar and, ultimately, comes under the operations team's control. As the operations team is bigger, older, and more powerful, it easily pressures the new team to disband innovative processes in order to comply with the existing procedures implemented for other applications and systems in the enterprise. Instead of becoming a core of innovation, the people in the new team become dissatisfied and leave. This can only be prevented by strong leadership, charismatic change agents in new teams, and attention from top management in the company.

The second approach is typically the "shift left" approach, which is executed on the enterprise scale and leads to the best-case scenario illustrated in Figure 1.2. In this case, the development team pushes the boundary to try to provide verified and tested release candidates to the operations team.

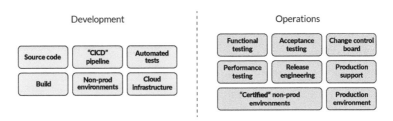

Figure 1.2. Model with expanded responsibilities for the development team.

Unfortunately, this can lead to duplicated effort, because the development team doesn't take into account the requirements of the operations team. The reason behind this is that the development team is not properly educated in IT operations and doesn't understand and follow the policies that the operations team work with. Without such understanding, any implementation of CICD pipelines and DevOps best practices becomes a facade, which resembles a proper change management pipeline in form but doesn't do enough in essence. In turn, the operations team cannot accept the CICD pipeline from the development team because their key policies have not been satisfied.

Even if there is an "appearance" of a CICD pipeline being implemented and the process being much more efficient, the reality is that the right-hand side of the process is still a bottleneck. After a certain point, it becomes impossible to shift more activities to the left-hand side without refactoring the process end-to-end.

1.3. HOW TO APPROACH TRANSFORMATION

There are many reasons why it is difficult to bridge the gap between development and operations. One of the main reasons is misaligned priorities between the two organizations, which is the result of a reporting structure where these two organizations are divided up to the very top of the enterprise. One of the first goals is to ensure that the priorities and key performance indicators (KPIs) between development and operations are aligned, which may require reorganization.

However, even when priorities are aligned, it is still difficult to bridge the gap for two reasons:

1. The operations team is attached to convenient processes, procedures, and tools that implement the required policies and controls.
2. The development team doesn't understand the policies that the operations team needs to comply with.

The root cause is that, over a long period of time, the policies and controls get mixed with the actual processes and procedures that implement them. The deep understanding of the actual policies, which those procedures implement, may easily be forgotten. The curious part is that, if we take a close look at the motivation behind the standards that ITIL, ITSM, and COBIT define, most of them make a lot of sense, even in the DevOps and CICD world. All of these standards and guidelines were written in the blood, or rather, the sweat, of operations engineers and are the result of many broken careers and lost opportunities.

Our approach to DevOps transformation is that we don't want to abandon anything blindly. But we do want to separate policies from processes and procedures, as shown in Figure 1.3.

Although most of the key policies are generic and repeatable from organization to organization, there is a certain degree of customization depending on the business, the organization, and even the applications that the organization is responsible for. For example, the management of Service Organization Control 2 (SOC2) certified applications and Payment Card Industry (PCI) certified applications may require different policies.

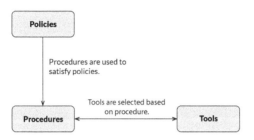

Policies are the constitution and laws. They are high-level, change rarely, and define core principles.

For example:
- Only authorized changes should be deployed to production.
- Every requested change to production should be approved by an operations lead.

Procedures are processes that define how policies are enforced. When the industry develops better workflows, these procedures should be updated accordingly.

Example:
- Change request should be created in Jira by a release engineer. An operations lead should review release notes and test reports and then approve the change. The change is then executed by a production operations engineer. After execution, it is reviewed by an operations lead to ensure that the change was applied successfully.

Figure 1.3. The separation of policies and procedures.

An IT organization typically knows what policies it needs to comply with. These policies may not be clearly documented, but the knowledge exists in the team and the minds of subject matter experts. It is, however, important to separate the policies from the procedures that the organization currently follows.

In the transformation to DevOps, policies become the common language between development and operations. It is much easier for developers to accept policies, because they define intrinsic requirements and are justified by the decades of experience of the operations team. It is also hard for anyone to argue that a company does not need to be SOC2 or PCI compliant. Unlike procedures, which offer solutions that quickly become old and outdated, policies define requirements and invite both development and operations to work toward a common solution.

During migration to cloud environments with microservices architecture, adoption of DevOps, and implementation of continuous delivery pipelines, the policies should remain the same, at least initially. Procedures, on the other hand, should be redesigned and a new set of tools should be chosen. Deeper changes in policies may also follow after the new procedures have started working, but it is best to do such reviews after the changes to procedures and tools. The reason is that changing procedures while retaining policies will bring the operations and development teams closer together, which will, in turn, start changing the culture.

To better understand the above approach, let us review an example. Let's say there is a policy that every change request to production should be approved by the development lead. An old fashioned procedure may require a change request to be created in Jira or ServiceNow referencing the specific version of an artifact that has to be deployed. Then, there is a step in the tool's workflow when the development and operations leads will approve that change request. If the code is stored in Git, a new procedure may consider a Git pull request to a source code repository as the change request, limit the code review approval to the leads, and consider an approval of the pull request to be an approval of the change request. That way, no additional creation of tickets is needed, there is less room for errors of providing wrong versions in the ticket, and the whole process is seamless. When taken in isolation, this may feel like an artificial example, but it is actually used by some companies to successfully pass formal audits and satisfies all required policies. In combination with other improvements, it may lead to an extremely lightweight process and help in shaping the right culture by bringing the developers closer to production operations.

1.4. ASSUMPTION AND CONSTRAINTS

Although this framework is generally applicable for a wide variety of situations, we are going to describe the case of cloud replatforming and service-oriented architecture in this book. Later in the book, we will give a specific example of an eCommerce platform transformation, but we have successfully used this approach for the transformation of companies in other verticals, technologies, and business domains. The same concepts will apply to other architectures, but the specific implementations will vary.

2

DEFINITIONS

There are many ambiguous terms in the industry related to continuous delivery. We want to stick with industry standard definitions, but we'll clarify exactly how we are going to use them in this book.

2.1. CLOUD AND MICROSERVICES

The term cloud [4] is used in the industry quite freely. In this book, when we refer to a cloud, we will mean infrastructure-as-a-service (IaaS). Specifically, all features of that infrastructure should be available via an application programing interface (API). If VMs and storage are provisioned by raising tickets in a self-service portal, we will not consider it as a cloud. Typically, when we talk about clouds, we will specifically mean cloud providers like AWS by Amazon, GCP by Google, or Azure by Microsoft. Most of the topics discussed in the book related to clouds can be applied to private clouds hosted on premises with VMWare and OpenStack, but, in our experience, these clouds are rarely implemented correctly and they typically lack functionality and stability in comparison with public cloud offerings.

A special disclaimer goes to platform-as-a-service (PaaS) clouds. If we want to talk about platform-as-a-service cloud offerings like Heroku, BeansTalk, and CloudFoundry, we will specifically mention PaaS.

Microservices [5] is a new approach to the architecture of complex applications. When we talk about microservices in this book, we will oftentimes mean just a good service-oriented architecture with relatively fine-grained services. The only real constraints that we'll talk about are that a single service cannot span team boundaries, teams should be sufficiently small (follow the Amazon "two pizza team" rule), and communication between services should be via a well-defined contract.

When we talk later about systems, applications, services, and components, we will mean the following [6]:

- Systems will mean large collections of services.

- Services will correlate, more or less, with the traditional definition of services or microservices. Services should implement some business-level functionality in a bounded context from domain-driven design. They may consist of one or many custom applications and system components and are available via a well-defined contract and API.

- Applications will usually mean business applications that implement some business domain logic, that are oftentimes custom applications, and that are sometimes available via a user interface (UI) and sometimes via an API.

- Components will refer to building blocks smaller than a service that are individually deployable and executable. An example of a component may be a database, a cache, or a custom business application.

2.2. DEVOPS AND CICD

In the industry, the term DevOps [7] is used in multiple ways: it can mean either an organizational culture, a set of specific processes, a team, a project role, or an engineering position. We stick with the definition of DevOps as a culture with a number of loosely defined signature best practices, the main of which is bridging the gap in skills and processes between the development, QA, and operations teams.

Continuous delivery [8, 9], on the other hand, is a specific set of processes aimed at decreasing the time-to-market and increasing the efficiency of development and operations teams, without sacrificing quality and controls. Sometimes, when people want to describe an end-to-end process, they talk about it as continuous integration, continuous delivery, and continuous deployment, to split the process into the phases of build and unit tests, quality assurance and other release activities, and production deployment, respectively. Continuous delivery may also be extended by an Agile project management process, like Scrum and Kanban, to cover the full application management lifecycle. Some people would define Agile as a broader process that encompasses continuous delivery, but it doesn't prescribe or require this.

In this book, we will assume that continuous delivery covers the process from writing code to production deployment. For brevity, we will oftentimes use the term CICD instead of continuous delivery. If we want to talk specifically about the continuous integration and continuous deployment phases of the pipeline, we'll call it out.

2.3. CHANGE MANAGEMENT

We will use the term change management [10] when we want to talk about the end-to-end and all-encompassing process of managing changes in production. By change management, we will mean not only the actual process of changing the production environment but the whole process from the business or other stakeholders requesting a change in the form of a requirement through to the developers implementing the change, the QA team testing it, and the operations team deploying it to the production environment.

Change management on the enterprise level typically includes other processes and types of changes, not only software-level changes. Traditionally, it deals with the actual application of changes to production and doesn't follow the changes end-to-end [11]. This may be one of the gaps between operations and development teams, because development teams talk about Agile and CICD end-to-end, but operations teams talk about change management for the change requests to deploy an artifact in production.

Ideally, we want to redefine the term change management so that both development and operations teams use it and it includes end-to-end changes from business requirements to production deployment. In fact, we will use the term change management a lot to describe this specific process.

When we talk about an abstract change, we will mean an end-to-end business change. For example, a business wants a new feature. This feature is a change, and this change needs to be managed. If we take software-related business requirements as an example, depending on the stage in the process, changes can take multiple forms, as illustrated in Figure 2.1.

One of the goals of end-to-end change management is to connect the changes from all of the points of view of all participants in the software development and release process. Another goal is to optimize the organization, architecture, and process for business changes and map production deployments to code commits and business changes as closely as possible. In an ideal world, every single business requirement would lead to one commit to source code, one new version of artifact, and one production deployment. That way, the scope of deployments is minimized, the risks associated with deployments are minimized, troubleshooting is made easier, and business-level A/B testing becomes easier to perform. Unfortunately, the building of such an ideal process may be prohibitively expensive, but we will show in this book how to get closer to it.

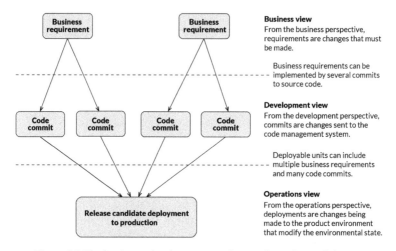

Figure 2.1. The business, development, and operations views of change.

A really good change management process should not only bridge gaps between development, QA, security, and operations, but should also include business stakeholders and product managers.

2.4. SITE RELIABILITY ENGINEERING

Site reliability engineering (SRE) [12] is a concept in which production support engineers have a deep understanding of the systems that they are supporting. An SRE team can do reactive and proactive tasks, including setting up and improving monitoring and logging systems, automating failover and recovery tasks to achieve self-healing, and fixing infrastructure, deployment, and application defects.

3

STARTING POINT

To begin with, we will consider an existing organization in an enterprise that is responsible for a portfolio of applications in a certain business domain. As an example, we'll use an eCommerce website, but this process is generalizable and can be extended to many other domains and architectures.

3.1. ORGANIZATION

Before we get to our example, let us give some historical overview of how IT organizations have evolved in big enterprises over the past several decades. We'll focus on traditional companies whose main business is not related to software.

IT organizations initially acquired and managed hardware and out-of-the-box software products from vendors. They started by managing internal data centers (first small ones and then bigger ones), desktops, and software systems, such as email servers, human resources tools, financial tools, and enterprise resource planning (ERP) tools.

IT organizations didn't need strong continuous delivery capabilities because custom development was limited and cost efficiency was the first priority. The responsibility to provide innovation fell on the software vendors, whereas the internal IT organizations just needed to manage the software products. With the movement to software-as-a-service (SaaS), some of the functions of IT organizations were outsourced. But the real need for internal transformation and innovation came when the enterprises needed to customize out-of-the-box software products and develop new functionality for themselves (Figure 3.1.). This challenge was amplified by the fact that more functions of companies were being digitalized and more companies started engaging with their customers through software and digital interfaces. We will take the retail domain as an example in this book precisely for this reason. Whereas traditional retailers in the 20th century engaged with customers in physical stores, retailers in the 21st century rely on the digital customer experience more and more.

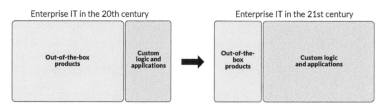

Figure 3.1. Shift from out-of-the-box products toward custom applications in the 21st century.

As more and more customizations were required, enterprises started to recognize that certain parts of their systems now had a direct impact on their business and, oftentimes, innovation in software was their differentiation from the competition. With that, the speed of development of new features became more important than just the efficiency of managing existing out-of-the-box products. In order to satisfy new business needs, new departments were created that were focused on software development. These departments oftentimes reported to a new executive-level reporting chain, and the position of Chief Technology Officer (CTO) was introduced or redefined to focus on information technology and digital capabilities.

Nowadays, the organization and reporting structure of a software development and delivery organization varies from company to company. There is, however, a typical set of roles and teams, and their relationships are similar across industries (Figure 3.2.).

From the top level, the infrastructure, development, testing, and other teams may report into two different lines: application development is reported to the CTO, whereas infrastructure and everything else goes to the Chief Information Officer (CIO) or Chief Operating Officer (COO). This dichotomy oftentimes leads to inefficiencies and challenges.

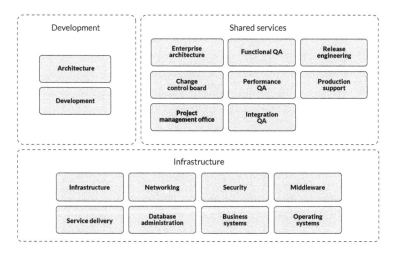

Figure 3.2. Typical roles and teams within software development and delivery organizations.

The infrastructure department is typically a traditional IT organization that manages out-of-the-box products and provides shared infrastructure services to application development teams. It typically has a number of teams:

1. Infrastructure provides servers to the application development teams.

2. Networking is responsible for the networking infrastructure, firewalls, and oftentimes load balancers.

3. Operating systems (OSs) is responsible for the base operating system administration and patching.

4. Middleware is responsible for the deployment and configuration of shared middleware components, like JVM, and application servers, like WebLogic or WebSphere.

5. Database administration (DBA) is responsible for traditional database and data warehouse administration.

6. Security is responsible for all aspects related to the security required to pass internal and external security audits.

7. Service delivery is responsible for operations aspects, including production support, incident management, and problem
management.

As a traditional IT organization, the infrastructure department
still manages various enterprise systems, which are typically out-
of-the-box products and don't require active development. Examples of such systems include ERP and email.

The software development department focuses on implementing
custom applications for the business. The roles and sizes of these
departments have been increasing lately. In general, the software
development department is broken down into multiple departments, each of which is responsible for a portfolio of applications
or a business domain. For example, in a retail commerce company,
these departments include:

1. An omnichannel digital customer experience platform, which
implements the front-office customer-facing services that are
exposed on web, mobile, social, and other channels.

2. A fulfillment platform, which implements the inventory and
order management services.

3. A business intelligence platform, which aggregates all data in
the enterprise to provide business reporting and insights.

The list above is not exhaustive but should give a sense of what
we are trying to describe.

Software development requires the collaboration of multiple departments and teams, including architecture, development, quality assurance, release engineering, production operations, project
management, and product management. Typically, these teams
have their own reporting chain and report to a Director or Vice
President (VP) level person, who in turn may report to the CTO
or CIO. The work on a project follows a matrix organization: when
required, personnel from each team are allocated to work on a specific task for an application or service.

Let's take a closer look at the typical roles involved in software
development, as outlined in Figure 3.3.

Software development departments are responsible for a portfolio of applications within a business domain and include several
development teams working on multiple applications. Applications

may vary in size, with big ones having several development teams assigned to work on them in parallel.

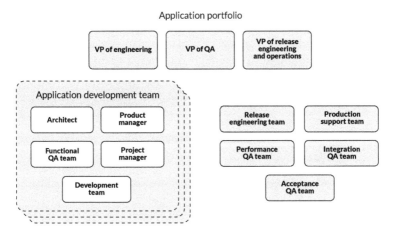

Figure 3.3. Typical roles in software development.

A software development department involves the following shared teams and roles:

1. The release engineering (RE) team is responsible for the change management process and manages the releases of applications to production. They work with all applications in the portfolio and the change control board (CCB) on the enterprise level to orchestrate changes across departments.

2. The production support team is responsible for supporting applications in production. This team deploys the changes in production prepared by the RE team. Depending on the company, a production support team that belongs to the software development organization may be responsible for end-to-end support from level 1 (L1) to L3, or they may be responsible for L2–L3 application-level support with L1 support performed by an enterprise-wide service delivery team.

3. The integration and acceptance QA teams are responsible for functional testing and certifying the quality of all applications in the portfolio.

4. The performance QA team is responsible for non-functional testing of all applications in the portfolio.

Application development teams, in turn, consist of the following teams and roles:

1. The product manager reports to the business and is responsible for defining the product roadmaps and business requirements for applications under development.

2. The program or project manager reports to the project management office and is responsible for managing the programs and projects that the department runs.

3. The architect reports either to enterprise architecture or to the VP of Engineering inside the department and is responsible for application architecture and compliance with enterprise-wide architecture standards for the technologies being used, security, etc.

4. The application-level functional QA team is responsible for functional testing of an application under development. Typically, each team is responsible for the quality of only one application.

5. The development team is responsible for implementing business requirements.

3.2. CHANGE MANAGEMENT

In most software development organizations as described above, the release engineering process evolved from the change management processes in traditional IT organizations. We will start by describing what types of changes need to be managed.

3.2.1. Changes

In order to understand the different types of changes, we need to analyze the different entities that can be changed in production (Table 3.1.). As we discussed before, we want to consider an end-to-end change management process that starts with business requirements and ends with production deployment. So, in order to be considered as a change, something needs to be connected with the business demand on one side and with an entity that is a part of the production environment on the other.

Table 3.1. *The conceptual types of changes that should be considered.*

Entity type	Drivers	Description
Applications	Business	Changes to business applications: when the business requests new functionality, the development team implement it in source code and the operations team deploy and manage it in production.
Application runtime properties	Operations	Changes to application runtime configuration that are required to maintain applications' service-level agreements (SLAs). Most often consist of configuration of external dependencies, capacity and sizes of clusters, and other technology-related settings.
Application business properties (feature flags)	Business	Dynamic configurable changes to application behavior in production. Typically implemented via feature flags and dynamically reconfigurable by business users.
Middleware	Business, operations	Changes to middleware components, as well as upgrades to new versions of middleware components and other out-of-the-box systems, like databases. Typically initiated to maintain application SLAs or address security concerns when a corresponding software vendor releases a new version. Can sometimes be indirectly initiated by business requirements if a new requested functionality cannot be satisfied with the existing middleware version.
Operating systems	Operations	Operating system upgrades and security patches. Typically required to maintain application SLAs and satisfy security requirements.

Entity type	Drivers	Description
Infrastructure	Operations	Base infrastructure changes, including addition of capacity to computer or storage infrastructure in the form of servers, VMs, or disks. Oftentimes required to maintain proper application SLAs.
Networking	Operations	Networking changes including network design, firewalls, and load balancers. Typically required to maintain application SLAs and sometimes to satisfy security requirements.

One of the common drivers for all types of changes is security, which may affect anything from application code changes to infrastructure changes.

There are several major items that are out of the scope of change management:

1. Tests and test data. Although tests and test data are very useful in non-production environments and should be versioned together with application code, they are not deployed to production. Therefore, they are not technically changes and are out of the scope of change management.

2. Production data. Changes in data that are generated during normal production operations affect the functionality of the production environment and may affect the production SLAs significantly, but they are not in the scope of change management. Applications that we deploy to production should be able to work with any configuration of data, and they should be tested to support this. Therefore, changes in production data are actually changes in production, but they should not be managed.

The view above is meant to represent the current state of change management. We will review and restructure the change types, as well as the "in and out of scope" changes, when we discuss the new change management processes that facilitate the continuous delivery, DevOps, and cloud and microservices deployment architectures.

3.2.2. Policies

No matter what the type of change, there is a typical set of policies that govern the change management process; these policies follow the generic rules that only authorized changes can be deployed to production and each change should be auditable. Policies are often defined to satisfy internal and external audit requirements. Traditionally, base change management policies were defined by ITIL and ITSM and were then customized to the company on a case-by-case basis. In this book, we will not stick with the formal definitions from ITIL or ITSM; instead, we will give a more conceptual overview of change management policies.

Let's start with the policies that define how to request a change. Only those people in a well-defined set of roles in the organization should be able to request a change. On a high-level, the different roles are indicated in Table 3.1. based on the drivers of the change. For example:

- A new business functionality can be requested by a business or a product manager. If a product manager delegates that role to a systems analyst and this person creates a requirement definition, the product manager should still review and approve the requirement.

- A change related to security should be requested by the security team.

- A change related to the improvement of application SLAs should be requested by the development or production support teams.

After a change is requested, it should be prepared or developed. In the case of a business change, this typically involves writing application source code. In the case of infrastructure changes, it can involve writing scripts. In all cases, preparation leads to the creation of an executable plan for how a change can be executed and applied to production. The plan includes various degrees of automation, but it ultimately represents an algorithm that can be executed by the production operations team. The critical step is that, after it is prepared, the change should be approved and signed off by the lead of the team responsible for preparation of the change.

Specific policies that define approval processes vary by team and type of change and include the following:

- A change should be peer reviewed by the senior members within the team.

- A change should be tested with the tools internal to the change development team.

- A change should be analyzed with code analysis tools defined within the team for defects and security violations.

After a change is developed, it should be tested before it is executed in production. This step is defined by policies requiring approvals from the respective quality assurance team leads. Quality assurance is a broad topic with a significant body of knowledge, so we'll cover it only briefly in this book and will only touch on those aspects that apply to change management. We will focus on a typical implementation of quality assurance that involves quality control and, specifically, testing. Typically, there are several types of testing that may be covered by one of a number of teams:

- Functional testing.

- Performance testing, which oftentimes includes stability and stress testing.

- Security testing.

- Acceptance testing.

Functional and performance testing can be further split into application-level and integration, feature-level or regression, etc. The list above can be extended to other types of testing, such as usability testing. The addition of more types of testing does not break the model, which can be extended to include as many other test types as required.

The common challenge behind testing is that it requires a properly configured non-production environment that is sufficiently similar to production. For every type of testing, we will define a number of additional policies:

- Tests should be performed in an environment that contains the change. This policy ensures that the tests are running in an environment with the required versions of the application and infrastructure deployed.

- The change should be deployed to the test environment in the same way that it will be deployed to the production environment. This policy ensures that the process of applying the change to the environment under test will be the same as the process of applying it to the production environment. This is especially important if the process of applying changes to environments is not idempotent.

- Test environments should be indistinguishable from production. This policy ensures that the system under test will be similar to production from a configuration perspective. It doesn't mean that the non-production environment should be exactly the same, because this is impossible. We just need to ensure that the tests do not distinguish the difference. Specifically:

- With regard to capacity, this means that, for functional testing, environments can be provisioned with lower capacity, but they should have a similar capacity to production for performance testing.

- For test data, we oftentimes can't use production data, so the test data that we use in the non-production environments should be sufficient to cover all combinations that are possible in production and should have sufficient size to test performance and scalability.

- From an external dependencies perspective, we should use similar instances to those in production. In cases when this is impossible, appropriate test endpoints of real dependencies should be provided or stubs should be implemented.

Sign off from the respective QA leads requires that tests have been performed on properly configured non-production environments, test results have been analyzed, and appropriate decisions have been made.

After the change has been tested and approved by the quality assurance and security teams, the change should be approved by the operations team from a maintainability perspective.

Finally, the change should be applied to production. After deployment to production, the production state should be reviewed by a person other than the one responsible for execution of the change, to ensure that the change was indeed applied in the right configuration.

All of the approvals, test reports, deployment logs, and environment configurations should be tied to a change. The full change lifecycle should be written in the audit log so that it will be possible to answer the following questions:

- When was the change deployed to production?

- Who verified that the change was successfully applied to production and the production environment was in the desired state after the change was applied?

- What release notes and production deployment logs are available?

- Who executed and approved deployment of that change to production from the operations team?

- Who approved the change from the functional, performance, and security perspectives?

- What were the test results, test analysis reports, and code and feature coverage metrics for this change?

- Who prepared and developed the change?

- Who initially requested and approved the change?

Figure 3.4. illustrates the policies (not the process) discussed above. To summarize, the key change management policies are:

1. A change can only be requested by authorized team members.

2. A change should be approved by the lead of the team that prepares and develops the change.

3. A developed change should be approved by the quality assurance lead.

4. A developed change should be approved by the performance testing lead.

5. A developed change should be approved by the requestor (the business user in our case).

6. A developed change should be approved by the security lead.

7. A developed change should be approved by the operations lead.

8. A developed change can be implemented in production only after all approvals have been collected.

9. A deployed change should be reviewed by the requestor, operations, and the quality assurance lead.

10. An audit log should be available for all changes end-to-end.

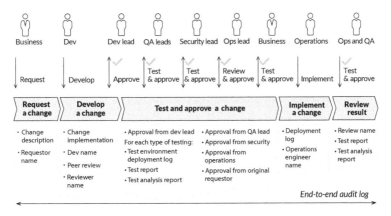

Figure 3.4. Summary of key policies governing the change management process.

Although some of the policies above look prescriptive, they can be implemented in a variety of ways. Some implementations may be extremely cumbersome and inefficient, whereas others are lightweight and efficient. One hint for a future discussion of efficiency is that Figure 3.4. and the previously discussed policies are based on roles. In practice, some people in the team can play multiple roles.

3.2.3. Procedures

Let's now consider how the above policies are typically implemented in enterprises. The teams that are responsible for implementation of these policies are typically release management and the CCB
[13]; release management is typically a team inside each software
development department, and the CCB is a corporate function in-
herited from traditional IT departments.

The release management team implements a number of processes and procedures to satisfy the policies defined above, as illustrated in Figure 3.5.

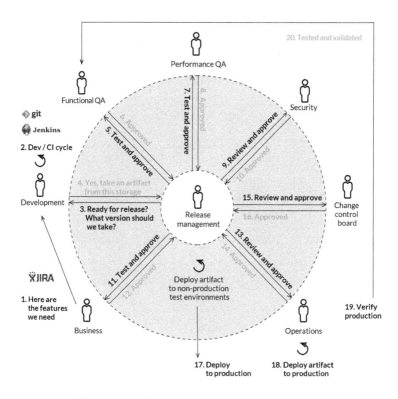

Figure 3.5. Processes and procedures of the release management team.

We will not describe the current processes and tools in detail
because they vary from company to company. The key point here

is that these processes rely on manual orchestration by a central body: release management. Although individual steps of the procedures may be automated, the whole process is still manual and often relies on significant ad hoc peer-to-peer communication. The procedures that teams implement to satisfy change management policies are often manual as well. This includes environment provisioning, application deployment and configuration, functional and performance testing, analysis of test results, and generation of release notes. The abundance of manual steps in the process can lead to major inefficiencies.

The end-to-end process becomes even more complex if one or more steps fail along the way. For example, if a QA lead does not approve a release because of the presence of critical defects, those defects need to be fixed and the process needs to be restarted.

3.3. CHALLENGES

The organization structure and processes above contain a number of challenges that are manifested in a long time-to-market cycle, high cost, and inefficiency of the process. It typically takes a long time to get a single change through the pipeline, it is difficult to get many changes through the pipeline at the same time, and each iteration of the process requires significant effort from the organization.

There are several primary root causes of these challenges that come from the organization, architecture, and process areas.

Let's consider the organizational issues first. The organization structure described above is not optimized for processing changes. A single change requires the effort and coordination of multiple teams. Such coordination is done via email, JIRA, or other semi-manual tools, which makes team collaboration very time-consuming and requires significant orchestration efforts from project managers and release managers.

Each team involved in the development of changes has its own backlog and the priorities of these teams are often not aligned on a tactical level, so the end-to-end process for a single change may take weeks or months. The only point where the reporting lines of the teams merge is at CEO level (Figure 3.6.). So the problems become even worse when things start breaking or one team starts blocking another. In order to get the process working, management needs to spend a lot of time on building internal alliances and playing political games; this sometimes works but is not generally a reliable delivery mechanism.

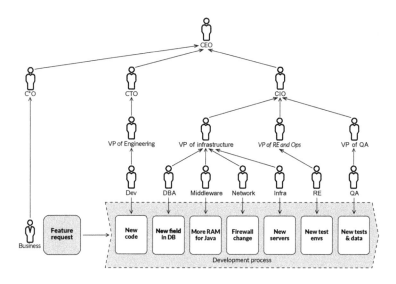

Figure 3.6. *Organization structure and reporting lines for change development.*

Our second major challenge is that the architecture may not be optimized for processing business changes. A single change may affect multiple components, which will take longer to implement and is more difficult to coordinate.

On the other hand, a single component may be so large that it requires multiple teams to work on it together, which complicates collaboration. Because multiple teams work on the same codebase, the development of changes is typically serialized. Even if fancy strategies, like feature branches, are used to parallelize work within a single application codebase, the application remains a single cohesive deployable unit and the work is still serialized. At the same time, excessive branching can lead to other issues, in the form of the difficult resolution of merging conflicts.

As can be seen from Figure 3.7., if there are three different features requested by different business departments, they cannot be developed in parallel. The orange and blue features will only be ready at time T2, whereas the green feature will only be ready at time T3. This happens because the features affect different applications and the work on each application is serialized as a result of the same teams working on the same codebase. Oftentimes, the application code is not suitable for partial feature releases, so once work is started, it needs to be finished, which makes the cycle even longer.

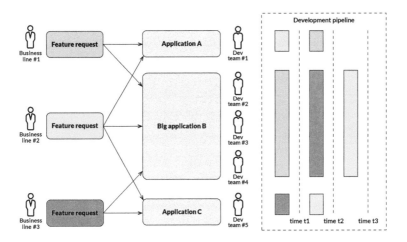

Figure 3.7. Serialization when multiple teams are working to develop new features for different applications.

One of the more subtle architectural concerns is that, even if the whole system is designed as a collection of services, there may not be well-established boundaries and contracts. The consequence of this flaw is that, even if a change is done in a service somewhere down the stack, that service cannot be released independently. If strong contracts are not in place, the whole system will need to be retested end-to-end to ensure correct functionality.

Another major challenge is that manual processes and long release cycles lead to long feedback loops for development teams. By the time the release fails in the final stages of testing because of a defect, the developers could have already developed more features and forgotten about earlier code changes. Fixing the defect and rerunning the whole pipeline again will take significant time and increase the overall time-to-market.

This problem can be seen in Figure 3.8. Three weeks after a feature was developed, it failed in the last stages of the testing process with two other new features. When the developers have recalled the feature and fixed the defect, do they need to release it with all of the features that have been implemented and partially tested during those three weeks? If yes, it will lead to more problems that we will discuss below. If no, there are potential ways to use the advanced techniques, like cherry-picking, of modern source code management systems like Git. Unfortunately, these advanced techniques are complex to use, more error prone and therefore risky, and difficult to reason about from the change management perspective, so they are not often used in practice.

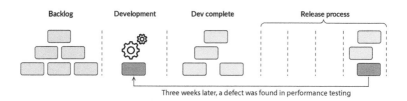

Figure 3.8. The feedback loop.

A long release process also leads to an accumulation of many changes in a state where they are developed and committed to the codebase but not yet released. This increases the scope and risk of production deployments, which, in turn, demands that more thorough testing should be performed. This creates a feedback loop and leads to even longer release cycles, ultimately completely paralyzing any deployments. In some cases, the whole process can choke and development teams need to be put on hold to allow all pending changes to be tested and released to production.

Finally, if the process consists of many manual steps and coordinations between the teams, audit logs are difficult to maintain and they are sometimes forgotten. It also gets hard to measure the effectiveness of the process and to understand where the bottlenecks are and eliminate them.

To summarize, if we consider a change management workflow as a service, it typically doesn't satisfy SLA requirements for either latency or throughput of changes that it can process, and it can sometimes choke if too many changes need to be processed. Furthermore, the system doesn't have necessary the logging and monitoring information to diagnose and fix issues.

3.4. TRANSFORMATION

The challenges above can be fought against by pouring in more money and hiring more people to support the process. Unfortunately, the problems can become very visible, especially when competitors are pushing companies to transform how they are doing IT and software development:

1. With increased competition and increased speed of innovation, business is asking technology departments to innovate and deliver features faster.

2. With data-center leases reaching their expiration dates and competitors moving to mature public cloud providers, executives are pushing their organizations to save operating costs and migrate to cloud infrastructure.

3. With maturing open source software and expiring contracts with software vendors, executives are pushing their teams to minimize license costs and migrate to open architecture stacks.

4. The solid best practices of microservices architecture, automation of various aspects of application lifecycles, DevOps and continuous delivery, and pressure from team members leave no excuses for not adopting new approaches in house and gaining the benefits that come with them.

To satisfy demand, software development and IT departments need to transform. The rest of this book is dedicated to working through such a transformation and providing a framework for how change management processes can be implemented to embrace clouds, microservices, automation, DevOps, and continuous delivery.

4

ARCHITECTURE

We'll start with architecture and analyze the new requirements and challenges that are facing modern architectures. The usual purpose of applications is to satisfy business requirements and to do so within predefined SLAs. The goal of architecture and design is to satisfy these requirements with the available technology stack, to provide enough extensibility and maintainability, to take into account security constraints, to comply with laws and industry standards, and to do so within the budget constraints.

4.1. ARCHITECTURE FOR CHANGE

The usual requirements for applications haven't gone anywhere. Software development departments still need to implement applications that satisfy business requirements and SLAs within budget. What has changed is the following:

- Business requirements have increased in number, complexity, and uniqueness from company to company with the increasing size and complexity of modern systems and applications.

- SLAs have become more strict. The number of users of applications has increased, those users expect faster response times, and downtime is not tolerated anymore. With the inability to grow computing power on a single server, applications now need to be horizontally scalable.

- Technology stacks have changed, with more mature, robust, and cheap open source technologies, which has led to migration away from expensive out-of-the-box applications.

- The establishment of cloud computing services with the pay-as-you-go model has led to requirements to implement dynamic scalability and fault tolerance for individual components that applications are hosted on.

To satisfy these requirements, large monolithic systems should be broken into multiple individual services to manage complexity and size. Each service should contain distributed, highly available, and horizontally scalable components that can tolerate failures of individual VMs, and the whole system should be able to tolerate failures of individual services.

In addition to the above requirements, business requires continuous innovation and a fast time-to-market of individual features for fast experimentation. The ability to change and adapt has become extremely important, which means that modern applications need to be designed with change management taken into account. We are not talking about generic extensibility anymore; applications need to support fast development of individual changes.

When a system is designed correctly from a change management perspective, as illustrated in Figure 4.1., each individual change should affect as few services as possible, the development of a new feature should involve as few teams as possible, services should be small enough so that a single team can work on them without overlapping or interfering with other teams, and services in the system should be loosely coupled and work over a well-defined contract to ensure that a change in one service's behavior does not negatively or unpredictably affect other services.

If an architecture satisfies the requirements above and the organization structure is planned accordingly, every change affects only a small part of the system and organization, which makes it easy to implement:

1. Teams being responsible for their own services means that a small and cohesive team can fully develop, test, and deploy the change to production.

2. Small services means that only one team is responsible for one service, so no orchestration of changes with other teams is required.

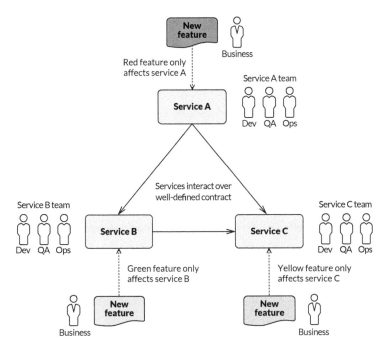

Figure 4.1. An ideal system from the change management perspective.

3. When services communicate over a well-defined contract, a change to each service can be tested and deployed in isolation, which makes the scope of change smaller and hence reduces the risk of deploying that change.

It is important to note that large business initiatives may still affect several services. However, a well-designed service-oriented architecture allows big features to be broken down into smaller ones, with each feature affecting only one service. When services' boundaries are chosen correctly, each small service-level feature remains meaningful from the business perspective.

In practice, it is often difficult to design the system in such a way that a single business feature always affects only a single service and a single team. The most obvious reason is that it is impossible to predict all future requirements. So, consider the exercise of creating such an architecture as an optimization problem. In order to approach this problem, assume that you need to minimize the total scope of change, which is the sum of all of the individual changes. The scope of each individual change is defined by how many services it affects from the development, QA, and deployment perspectives. Minimization can be achieved by splitting the entire system into a set of services. In order to do that, you should investigate how business teams and domains are structured, where new feature requests are coming from, and what the most typical feature requests are. This exercise can be done with historical data by assuming that the company's business is not changing and the business team structure is not changing dramatically. We hope that, in the future, machine learning can help with tackling this optimization problem more effectively in semi-automated way.

4.2. MICROSERVICES VERSUS MONOLITHS

When speaking about microservices, there are debates about how small you want to go. There are two extremes - a monolithic application at one end and tiny nanoservices at the other end (Figure 4.2.).

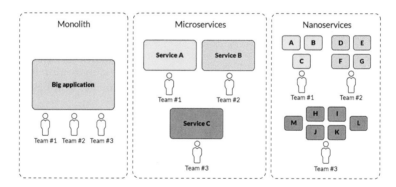

Figure 4.2. *Different types of architecture.*

A monolithic architecture is when all of the functionality and outcome from a software development department is concentrated into one "big application" consisting of a single component. The application in this case is very large and has multiple teams working on it. The benefit of a monolith is that the deployment architecture and testing approach is simple:

1. There is no complex interaction between the components of the application from the networking, routing, and load-balancing perspective.

2. There is no need to manage backward compatibility and dependency management in runtime.

3. When a new feature is implemented, the end-to-end functionality of the application is tested, so there is no need for separate service-level and integration testing.

4. Testing of the whole application for every change ensures that there will be no integration issues, which lowers the risk.

Unfortunately, monolithic architectures create the challenges for change management that we discussed earlier. There are examples of efficient change management processes with monoliths [14], but in our opinion, such examples are rare and require special conditions, such as the engineers, culture, processes, and technology stack all being mature.

In the case of monoliths, an application is internally split into modules from the design standpoint to manage the inherent complexity of business logic implementation. Engineers can wrongly assume that modules are the same as services, which is not the case because of the absence of well-defined enforceable contracts and tight coupling from the business logic, performance, failure domains, and deployment perspectives. This happens for the following reasons:

1. From a functional perspective, the amount of time that goes into the careful design of modules is often less than that in the case of services. So, in our experience, the boundaries of modules are typically less defined and the contracts between modules are not enforced well.

2. From a performance perspective, if one module starts performing poorly and starts consuming too many resources, it affects the other modules that are running in the same OS process and share resources with the offending process.

3. From an availability and failure domains perspective, if one module crashes, it typically crashes the whole system process with all other modules.

4. From a development perspective, working with a monolith with a large codebase may overload the integrated development environment, increase build times, and reduce developers' productivity.

5. From a deployment perspective, if only one module changes, the whole application still needs to be redeployed, which increases the scope and time of testing, scope and time of deployment, and risks associated with the change. In other words, there is no way to guarantee that the change affects only one module and, hence, that it has low risk.

The issues described above manifest themselves to different extents with different technology stacks. For example, if there is a web application that is implemented in HTML, CSS, or JavaScript, with all business logic residing on the client side within the web pages themselves, then some of the issues above can be mitigated with the right structure. Individual pages can be developed by individual developers, tested in isolation, and deployed individually without touching the whole codebase. But if the whole codebase is in Java or C, then not only will all of the issues apply but there will be additional issues related to long build times. Even if the technology supports modularity, mitigation of the issues above requires a very good development, testing, and operations team culture, highly skilled staff, and very good internal processes. Although we recognize that this combination of factors is possible, we never really saw it fully working in practice. So, such assumptions should not be made by most enterprise-level companies.

The other extreme is going too small with the size of services. In this case, the services themselves are very simple, but the complexity is shifted toward the "space between services," which will make the whole system more complex to understand, test, reason about, and change. Another concern with going to nanoservices is that, in order to ensure that the end-to-end business logic works well, tests will still need to be written on a set of the services, which increases the practical scope of each change. During one of our implementations, while arguing about the proper size of services, we had a joke that each service in an "ideal" microservices architecture should only return either "0" or "1" and we should let integration logic deal with the right order of invocation of microservices.

From the change management perspective, it is best to stick with the middle ground, where a single team is responsible for approximately one service and where a single service implements a reasonable and well-defined set of features. There may be practical reasons why you want to go with smaller services: they may better represent the business domain, it may be easier to plug in certain pieces of functionality and experiment with different versions of behavior, etc. But after a certain point of scale, an increase in the number of services will not improve the efficiency of change management any more. From the change management perspective, services' boundaries should be aligned with team boundaries; that is, a single team can be responsible for multiple services, but a single service should be developed by a single team.

4.3. EXAMPLE: ECOMMERCE PLATFORM

Let's consider an architecture for an eCommerce platform that provides a customer experience over a web channel.

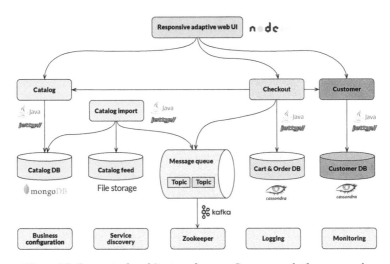

Figure 4.3. Conceptual architecture for our eCommerce platform example.

Figure 4.3 shows a conceptual architecture, in which each service may consist of one or multiple components and is colored with a different color. Gray-colored components are shared or infrastructure services. The architecture shown is by no means an exhaustive list of the services that are required for a typical eCommerce platform, but it contains a number of representative change management use cases that we will describe and investigate in detail later.

The technologies for this architecture are chosen more or less arbitrarily to include a representative list of popular technologies that are used in modern cloud applications. We will not discuss all possible options, because the focus of this book is not to evaluate modern open source technology stacks and create ideal software architectures for eCommerce but to discuss how to implement systems from the deployment and change management perspective. All of the approaches that we describe later will be applicable for other modern software development technologies. Technologies that are relevant for cloud deployments and change management, like service discovery and business configuration, will be discussed later and a number of alternatives will be given.

The architecture is split into the services shown in Figure 4.3. to represent the business domains of the whole application and to optimize for change management on the basis of typical feature requests. This means that new features will require changes in either the catalog, checkout, or customer domains. Integration with other systems is not described but does have to be taken into account from the requirements standpoint.

As shown in Figure 4.3., the web UI application is a standalone component. To further optimize for changes, the web UI component could have been split by business domains as well, so that each business domain would be vertically integrated. In this case, there would have been separate web UI components for the catalog, customer, and checkout domains. However, this was not done because of the difference in skill sets between the back-end and front-end teams (Javascript and Java) and to optimize for changes that affect the entire customer experience across domains. This type of architecture is most common, and we'll stick with it for the remainder of this book, to be more relevant to readers.

The responsive adaptive web UI service is responsible for rendering web pages on the server side. Client-side business logic is not considered here, and it is assumed that the JavaScript running in the browser works with the APIs provided by the web UI layer or uses APIs provided by the business services directly.

In our example, there is no need for a separate and complex API orchestration layer, because the number of services is low, they contain all necessary business logic, and the top-level services themselves implement the necessary orchestration if internal orchestration is required. For example, the checkout service makes the necessary calls to the catalog service to get information about products being purchased and to the customer service to get necessary shipping and billing information. An alternative approach would be to implement a separate API service that was responsible for complex orchestration of the services underneath. This approach would work, but it would make change management more complex because every change will inevitably touch the API orchestration layer in addition to the web UI service.

The catalog service implements functionality that allows customers to navigate through the products being sold on the website by using search, category browse, and product information pages. It exposes relevant methods that are used by the web UI and checkout service. The catalog service consists of a number of components:

1. The catalog application exposes the business functionality discussed above via a RESTful interface. This is implemented in Java and running on a Jetty application server.

2. The catalog database stores information about categories, products, stock-keeping units (SKUs), and their attributes. This is implemented in MongoDB.

3. The catalog import application listens to incremental updates in Kafka and updates the catalog DB in real time. This is implemented in Java and running on a Jetty application server. If lost, all data in the catalog DB can be replaced by running a full refresh logic in the catalog import application.

4. A number of topics in the message queue are dedicated for catalog real-time updates. The message queue itself is shared between services.

5. The catalog feed is a non-real-time method of refreshing the whole catalog on a regular basis. The feed is stored on the file storage, and updates to files on the file storage trigger a full refresh logic in the catalog import application.

The customer service encapsulates customer profile information and all methods of working with that profile, including the handling of basic customer account information, such as names and shipping and billing addresses, as well as payment information. As the customer service handles personally identifiable information (PII) and

PCI data, it has special security requirements. This service consists of two components:

1. The customer application exposes the business functionality discussed above via a RESTful interface. It is implemented in Java and running on a Jetty application server.

2. The customer DB stores information about the customers. It is a system of record for customer information, which means that the data cannot be replaced if lost. It is implemented in Cassandra.

The checkout service allows customers to create shopping carts and add and remove products in the cart; it also provides checkout and order management functionality. The checkout service internally uses the catalog service to retrieve information about products and SKUs that are in the cart, and it uses the customer service to retrieve information about the shipping address and payment methods during the checkout process. The checkout service consists of the following components:

1. The checkout application exposes the business functionality discussed above via a RESTful interface. It is implemented in Java and running on a Jetty application server.

2. The cart and order DB stores information about customer shopping carts and places orders. It is a system of record for both cart and order information, which means that the data cannot be replaced if lost. It is implemented in Cassandra.

3. The checkout service publishes messages to a topic in the message queue for order processing and fulfillment in external systems. It uses a shared Kafka message queue for this purpose.

We'll discuss the infrastructure components and dive more deeply into the deployment architecture of the above services in the next chapter.

5

INFRASTRUCTURE

One of the crucial aspects in the design of systems for continuous delivery is the right separation between infrastructure and applications. As we discussed previously, traditional infrastructure was managed by standalone IT teams and all interaction between the software development and infrastructure departments was manual (Figure 5.1.). This meant that when an application needed a new server, a change in networking or firewalls, or a new version of an operating system, a manual request to the infrastructure team had to be made, which typically involved a ServiceNow request, a Jira ticket, an email, or some other form of person-to-person communication. What made things worse was that all aspects related to middleware software and databases were handled in the same traditional way by the same IT departments. This significantly slowed down feature delivery, because many changes required the orchestration of work by multiple teams belonging to multiple departments.

To overcome these issues, infrastructure needs to provide services that are exposed via APIs with well-defined contracts and SLAs (Figure 5.1.). The easiest way to implement this is to use one of the cloud offerings from mature vendors like Amazon, Google, or Microsoft. Unfortunately, in our experience, private clouds like OpenStack and VMWare still lack the ease of access, stability, and security of the mature cloud vendors. It is very important to note that, when an IT department starts using IaaS, the cloud APIs need to be exposed to other departments. One of the infamous anti-patterns of cloud adoption is to expose IaaS services via a manual ticketing interface like ServiceNow. This method may be familiar for the IT department, but we believe it should be avoided, because it reinforces manual work and does not help to establish an efficient change management process on the large scale.

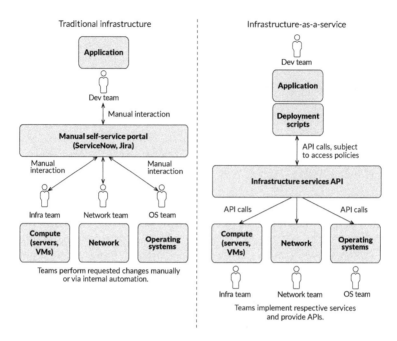

Figure 5.1. Traditional infrastructure versus IaaS.

We are going to assume that a company has a single infrastructure department providing "multi-tenant" services via APIs to all software development departments. The APIs that software development departments use to work with infrastructure services may be the cloud vendors' native APIs or separate out-of-the box or custom tools that provide a facade API for the cloud vendors' native APIs. It is important to note that reliance on public cloud vendors does not diminish the role of the internal infrastructure department. We have participated in projects where the infrastructure department felt that way, and it led to misuse of the cloud infrastructure because of a lack of proper experience within the development teams. Even if software development departments use native cloud APIs, the establishment of contracts with cloud vendors and the configuration of APIs is still performed by the infrastructure department. This is done for enforcement of enterprise-wide IT policies, like cost management, security, and access control.

When we discuss generic policies and ways to control them later in this chapter, we will sometimes refer to specific approaches that cloud vendors recommend to implement these policies and we will provide references to documentation in some cases. However, we recognize that some approaches that are recommended at the time of writing of this book will become outdated. So, we recommend that you do your own research and refer to specific cloud vendor documentation and training courses.

5.1. ACCESS CONTROL

First and foremost, the infrastructure team needs to establish proper access control to the IaaS API. The ability to manage access to cloud resources is an important security measure in itself, and it is also a prerequisite to implementing any other infrastructure policies.

Most cloud providers have in place robust access control mechanisms, which are commonly known as identity and access management [15–17]. The recommended way to implement top-level access control in all cloud providers is to create dedicated "projects" [18] or "accounts" [19] for different software development departments or teams and to have a dedicated project for the centralized access management and financial reporting that the infrastructure team will control [20, 21]. The granularity of projects and specific access control may vary, but the key point is that only the infrastructure team has administrative access and they are responsible for managing the access of other departments.

Personnel in software development departments and specific teams should be provided with access accounts with appropriate access (Figure 5.2.). These accounts typically come in two flavors: accounts for specific people and service accounts. Whether to use service accounts or not for automation that software development departments will implement is up for debate, but service accounts are used most often because of their ease of use. If service accounts are used, the responsibility to store them securely falls on the team that owns them. We recommend the following best practices to properly manage service accounts:

- A service account should have fewer or the same privileges as the individual people in the team that owns the service account.

- The account credentials should be stored in a secure place where only the members of that specific team have access.

- The automation tools or scripts should be parameterized to use any accounts (service or personal).

- If the higher-level automation scripts expose a service and API for higher-level consumers, they should provide separate authentication and access control as well.

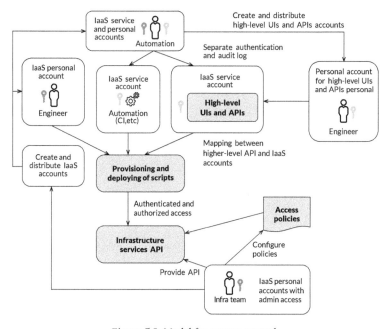

Figure 5.2. Model for access control.

Implementation of the best practices described above typically alleviates the security risks related to service accounts and provides a necessary audit trail. In any case, in order to ensure that proper security controls are established, the infrastructure or security team should establish audits of the teams owning service accounts.

We will come back to the topic of access control when we discuss other policies in detail.

5.2. COST POLICIES

Traditional IT departments are responsible for the cost of infrastructure. With migration to the cloud, this typically continues to be the case, so the infrastructure team needs to maintain certain controls to manage the cost of resources that the software development department utilizes. The management of these costs requires the infrastructure team to set up appropriate metering tools that allow costs of resources to be distinguished according to

1. software development departments,
2. teams within those departments,
3. applications and components that are developed by those teams,
4. various production and test environments that these applications are deployed to,
5. and specific people, owners, and requestors of these environments.

The list above gives only an example of the dimensions by which the cost metrics need to be collected and aggregated.

There are several methods to implement such cost metrics, and a combination of approaches is typically used. One of them is to implement tagging and labeling of provisioned resources in the cloud [22, 23]. Another is to separate departments, teams, and applications into dedicated cloud projects, which is the highest unit of segregation and isolation that cloud vendors provide. Once the cost is measured, it should be possible to control it. This is done by allocating infrastructure budgets at the department, team, or application level, establishing a chargeback mechanism, and enforcing those budgets.

The splitting of different departments and teams into separate cloud projects is the easiest option and can be accomplished by setting up proper access control, so that each department has access only to their project. Cost metering and quotas can then be set up for each project. Enforcement of proper tagging and labeling is more difficult, because the labels are set on resources during provisioning by the software development teams, not by the infrastructure team. In our experience, this can be done in several ways. The easiest way is for the infrastructure team to establish a naming convention, communicate it to the appropriate departments and let them use the proper tags and labels in their automation scripts. Violation of the agreement can be enforced with monitoring and during aggregation of costs by the infrastructure team. If some resources turn out to be improperly labeled (either an absence of a label or an incorrect name), the violation can be reported back to the appropriate team. The downside of this process is that it requires human intervention and quality assurance. In a typical company, we can assume that teams will do their best to comply and there is no malicious intent; hence, violations are rare and the inefficiency is acceptable.

A more strict but expensive approach is to implement a custom API as a facade and orchestration layer for the native cloud APIs. Tools like Cisco CliQr can provide a starting point for implementation of such an API facade. This API facade can then enforce tags and labels when resources are requested and can deny requests in cases of violation. A custom API can also set certain tags and labels by itself based on the authentication information without requiring any additional conventions between the infrastructure and development teams.

5.3. SECURITY POLICIES

The next major set of policies is related to security. We have already covered access control; in this section, we will discuss security policies related to networking and operating systems. It is important to note at this stage that, with the adoption of serverless architectures, OS-related security policies may be determined by cloud vendors and completely taken out of the scope of the client companies' infrastructure departments. However, at the time of writing, the task of keeping the operating systems of either VMs or containers secure and up to date is still typically the responsibility of an internal infrastructure team, so we will keep it within scope. Having said that, OS-related policies are typically relatively easy to implement by ensuring that the infrastructure team controls the list of operating systems available for provision by software development departments and that there is a procedure to monitor current versions of OS components and upgrade them in time when needed. The actual upgrade procedure can be implemented with two main approaches: either reprovisioning a VM or container as a whole if an update is needed or patching an existing VM at run time. Although the first immutable approach is preferred, we recognize that patching may still be useful in some cases. We will talk about the specific procedures in detail in the section related to Infrastructure Changes later in the book.

If an organization decides to use VM-based IaaS, OS-related security policies also involve OS-level access control. In the case of *nix operating systems, OS-level access control determines who can secure shell (SSH) into the VMs and what level of access they will have.

From the authentication perspective, there are several primary mechanisms, with the Lightweight Directory Access Protocol (LDAP) being the most popular in enterprises. On a smaller scale, authentication can be managed with SSH keys. Cloud providers may also provide convenient methods of managing SSH keys for access to VMs that can be configured on a project level [24]. Even if this is not the case, traditional configuration management tools can be used to update SSH keys on VMs, although this would require patching and violate the immutable infrastructure approach.

From the authorization perspective, LDAP is also the most popular choice. Oftentimes, root access is reserved for the OS team, and application teams are granted restricted access that is sufficient to deploy and run most applications. All settings are already baked into the available images that the application teams can provision.

Networking-related policies are typically harder to implement in the right way while still giving a good degree of automation and autonomy to the software development teams. There are several reasons behind this. The core reason is that a networking configuration is never isolated to a single software development team. The application that one team is implementing needs to communicate with other applications. Such internal communication cannot be done over the Internet for SLA and security reasons, which means that this application will be sitting on a private network. Private networks often span the whole enterprise infrastructure between clouds and private data centers. Another reason is that firewalls between subnets may need to be very strict to comply with internal or external security requirements, and the implementation of strict firewalls means relying on specific static sets of IP addresses where external dependencies are located. And finally, when it comes to private data centers, networking is oftentimes not automated there. The above reasons mean that software development departments cannot just create arbitrary networks or conFigure firewalls for them. The capacity of these networks and their address spaces have to be pre-allocated and pre-agreed to ensure that they don't intersect with the rest of the applications.

Having said all of the above, let's review one of the specific and most common cases of network-level security policies: firewalls. Firewall rules are typically defined by application types. In our experience, they can be split into the following categories by the communication that an application requires:

1. Outside world to application. This is required for Internet-facing applications. The firewall rules need to be configured on the target application side to limit what inbound connections

can be established and, if possible, from what IP ranges in the Internet. For example, if there is a secure web server behind Akamai, there will be firewall rules to allow inbound connections only on port 443 (HTTPS) and only from Akamai IP ranges.

2. Application to outside world. This is required for applications that need to communicate with other services that are available on the Internet. The firewall rules need to be configured on the target application side to limit external egress traffic from the application, for example, so that a rogue code in a credit card processing application does not transfer this information to malicious services outside of the enterprise. Often such egress limitations require the communication to go via a proxy, which, in addition to transport-level firewall rules, does application-level introspection and validation of the requests' contents.

3. Other applications to target application. This is required for internal applications to limit the security perimeter and mitigate security risks. Typically, the firewall rules need to be configured on both ends of the networks, where the client application resides and where the service application resides. Sometimes this communication also requires application-level, such as HTTP-level, proxies.

4. Target application to other applications. This is essentially point 3, but in reverse.

The reason we split firewall rules into these categories is that the scope of effect of such rules, and hence, the scope of changes to the rules, is limited to only one application or to several applications. As we discussed in the Architecture chapter, we try to design applications in such a way that each change should affect only one application. Unfortunately, in the case of networks and, even more specifically, firewalls, this is not the case. Although the first two types of firewall rules can be configured only for a single target application, the last two points also require configuration of the firewalls for other applications, which may or may not be in the cloud and may or may not be covered by automation; this makes configuration of these rules more static in nature. Obviously, this complicates change management and decreases productivity and efficiency.

One of the ways to mitigate the challenges with last two types of firewall configuration is to implement security policies on the level above the transport network level. For example, a policy can be implemented on the application side with authentication and authorization or requests. This is a valid approach that, unfortunately, is still rarely used in our experience. The practical difficulty with the implementation of such an approach is the lack of enforcement that the security team can establish for application development teams to ensure that their application-level access control is configured appropriately. Another problem may be related to prescriptive, rather than descriptive, external compliance requirements, like PCI. If both difficulties can be avoided, we recommend trying this approach because it will simplify change management. This is why most security and networking teams still prefer to work with firewalls.

For the configuration of all types of firewall rules in the scope of a particular application, there are three different implementation options:

1. Full access for configuring firewall rules can be given to the application development teams. In this case, the security and networking team will need to define the policies and requirements that an application should implement in their automation. Because the security and networking team are still accountable for setting up proper firewall rules, they will need to establish a robust audit process for the application teams. Depending on the skills available in the application teams and the specific security requirements, this may be too expensive to implement and may still not provide the necessary assurance that the security standards are being followed.

2. Firewalls can be defined and implemented for each specific application by the networking and security team. No matter how exactly this is implemented on the cloud API side, in this case, every time a change in the firewall is needed, the networking team will need to be involved. Let's say that the firewall rules are managed by setting special tags on the VMs. In this case, the networking team will set a rule that, by default, if there are no tags on the VMs, all communications are closed, and a set of tags will be created for each application with the specific firewall configuration. Even if a tag can be set on the VMs via the API and the application team are able to do this without involving the networking team, any change in the firewall definition will require human communication between teams. Unfortunately, even in this case, because the application teams have

the ability to set any tag from the allowed tags on the VMs, there should be some enforcement and auditing to ensure that the application team sets the right tags on the right VMs. The scope of this audit can be smaller than in the first case though.

3. An approach similar to that described in 2 can be used but with the firewalls created by application type, not by application. For example, there can be an application type "web server" that can have ports 80 and 443 open and an application type "secure web server" that can have only port 443 open. In this case, abstract tags are created for each application type, and the application teams can set any tags from the allowed ones for their applications. This case will still require collaboration between the networking and application teams and regular audits, but it may limit the number of cases when human interaction is required.

Depending on the situation, security requirements, severity of security risks for certain applications, and availability of skills in the application teams, different approaches can be used. In most cases, when considering migration from a traditional enterprise structure with the IT team having security and networking expertise and the application teams not having it, we recommend going with approaches 2 or 3. Although approach 2 is not the most efficient in the long run, it is the most straightforward one to implement in our experience, because it doesn't require the robust design of generic application types and their associated firewall rules.

In the longer term, application teams should be educated in security and networking, and approaches like 3 or 1 should be implemented.

5.4. OTHER POLICIES

Other types of infrastructure-related policies can come from the enterprise architecture: for example, which OS versions are allowed and what type of servers and storage can be used. In these cases, such policies are enforced by configuring proper access control on the cloud API side to allow provisioning of only allowed resources in allowed configurations. Mature cloud vendors provide very robust access management that can be tuned in a fine-grained way.

5.5. PRACTICAL ENTERPRISE IAAS

Although ideal implementation of IaaS would replace any human interactions with the service API between the infrastructure and development teams, it is not always feasible in practice. There are some enterprise-level policies that the infrastructure, networking, and security teams are asked to implement in a centralized way. Therefore, the application teams cannot have full autonomy in how they use the infrastructure. One approach to mitigate this problem would be to understand the policies that the infrastructure teams need to implement, let the development teams implement them, and have an internal audit process to enforce them. However, this approach is expensive, because development teams don't have the required skills to implement all of the policies and they will essentially need to have infrastructure subteams inside them. Implementation of such internal audit processes is also expensive, and some policies are hard to enforce by auditing.

On the other hand, certain components of infrastructure are just harder to automate and expose via APIs: for example, the creation of accounts and projects in cloud environments, or the creation and configuration of private networks that are extensions of the existing enterprise network, which is, in turn, not virtualized and not automated.

Most importantly, oftentimes, automation of these activities and exposure of them via APIs is not practical and cost effective, because changes are rare. Practical implementation of IaaS in the enterprise can still leave some aspects of infrastructure management that involve human interaction and sometimes manual work.

The idea behind a practical IaaS implementation (Figure 5.3.) is that resources that change frequently, together with new application features, are available via an API. Resources that change rarely or only once can be set up through traditional human collaboration with the infrastructure teams. When the infrastructure teams need to follow certain enterprise-wide policies, they enforce these policies either via access control and other restrictions on the API or with conventions and audits for the development teams.

The following resources are usually created manually through human collaboration:

1. Cloud projects, which are top-level resource holders provided by cloud vendors. Their creation often requires collaboration between the infrastructure and development teams to

understand capacity requirements and define resource allocation quotas and limits. The creation of cloud projects may also
involve the configuration of financial metering and reporting
by the infrastructure team and an agreement on what additional tags and labels to set on resources to improve the granularity of budget quotas and financial reporting.

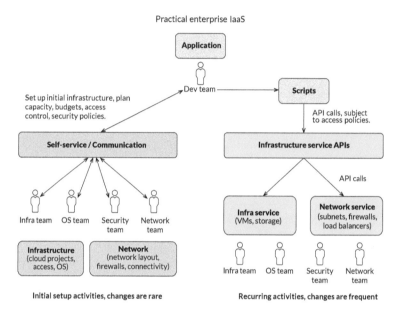

Figure 5.3. Practical implementation of IaaS involves some automation and
some human collaboration.

2. Access control, including the setup of personal and service
    accounts. This involves discussions between the infrastruc-
    ture and development teams to understand the roles within
    the team, service accounts that are required, and policies that
    the development team should follow to protect the service ac-
    counts.

3. Inter-data-center connectivity if a new cloud vendor is on-
    boarded or new cloud region is adopted. This work requires
    significant manual effort by the networking team to ensure
    that the applications are properly connected with other cloud
    vendors, private data centers, and other cloud regions. This

setup requires discussions with the development team about SLAs and non-functional requirements relating to the connectivity of their application with other applications. For example, if the development team is working on an application that should work with another application in a private data center located on the East Coast, this new application will need to be set up in the eastern region to ensure high performance and SLAs. Depending on the SLAs and security requirements, connection over Internet, virtual private network (VPN), or some form of Direct Connect will be used.

4. Base network setup and firewalls on the enterprise side. If a new application needs to be part of a private enterprise network, this new network should be designed, certain address space should be allocated, proper routing tables should be set up, and if the application needs to communicate with other applications in private data centers, firewalls should be set up on the private data center side.

5. Requirements for operating systems and OS-level access on the cloud side should be discussed between the infrastructure and application development teams. If an existing implementation of IaaS doesn't support the operating systems required by the development team, new operating systems need to be set up. If the development team requires root access or other non-standard requirements to VMs on the OS level, this should be discussed and certain activities should be performed by the OS team.

6. Network security and firewall setup in the cloud should be discussed between the teams as well. Based on the type of application, security policies, and development team capabilities, more or fewer permissions can be given to the application development team to conFigure firewalls in the cloud.

Although the list above may sound scary, most of the work is done only once during the initial setup and additional work is required only in unusual cases. This list should not be followed every time a development team decides to create a new service. Rather, it will be required when a new software development department is transitioned to the continuous delivery and cloud model. In this case, collaboration is required anyway to ensure that the department understands all of the implications related to the migration and also to ensure that certain education about infrastructure-level policies is performed. The most inconvenient parts of the list above are budget quotas, capacity limits, and firewalls. These resources may change from time to time, but practically speaking, changes to these are rarely required for new feature requests from a business.

All resources that are dynamic can be managed via an API that the development team can use:

1. New computing and storage capacity can be allocated and deallocated within limits via an API.

2. Load balancers can be created and destroyed on demand via an API.

3. In some cases, subnets and firewalls on the application side can also be configured via an API by the development teams, provided that conventions have been established and are followed between the networking and application teams.

5.6. A WORD ABOUT PAAS AND CONTAINERS

We haven't touched on middleware and databases yet in this Infrastructure chapter, although we mentioned that, in the current state, many companies may have those teams as a part of the infrastructure department. Unfortunately, in our experience, this rarely works well when transitioning to the continuous delivery and DevOps model. Middleware, like application servers, and system components, like databases, caches, and message queues, are integral parts of the corresponding business services. The exact technologies to use depend on the data storage and access patterns, which in turn depend on business requirements and are service specific. The best team to choose the right tools and conFigure them properly is the application development team, not the shared infrastructure team. The only legitimate time when middleware and databases can remain part of an infrastructure service is when they are provided as services via an API to software development departments. This would mean that the infrastructure department provides a PaaS, in addition to the IaaS.

We will discuss the challenges related to PaaS later in this section. Before we get to PaaS, let's talk about containers and container management platforms, because they may be confused with PaaS. Examples of container management platforms are Kubernetes,

Mesos, and Docker Swarm. Although containers have revolutionized the way that applications are packaged and deployed, they should be thought of as lightweight VMs and their management platforms should be considered as lightweight IaaS, not PaaS. Table 5.1. briefly compares traditional VMs and cloud services with containers and container management platforms.

Table 5.1. Comparison of traditional VMs and cloud services with containers and container management platforms.

Function	VMs and IaaS	Containers and "Container-as-a-service" (CaaS)
Packaging and deployment unit	VM image, stored in cloud provider storage.	Container, stored in Docker Registry.
Packaging toolset	Hashicorp Packer, Spinnaker, custom scripts using cloud APIs.	Dockerfiles.
Scheduling (resource management and placement)	Cloud IaaS computer API supports provisioning of VMs from images.	Kubernetes, Mesos, Docker Swarm.
Lifecycle management (auto-scaling, self-healing)	All cloud providers support auto-scaling and self-healing via an API. For example, AWS auto scaling groups and Google instance groups.	All container management tools support this via their mechanisms. For example, Kubernetes replication controllers.
Networking	Native and robust features including network and subnet configuration, firewalls, and load balancers.	Different container management platforms support this to different degrees. For example, Kubernetes supports overlay networks and internal IPTables-based load balancers, and it integrates with cloud load balancers.

In addition to the similarities described above, containers and container management platforms provide additional benefits:

1. Containers are more lightweight in build time. They don't have the unnecessary packages that traditional VMs come with, because VMs have fully fledged operating systems; this means that they can be patched more rarely and allows for

implementation of a true immutable infrastructure. Containers are also more lightweight in runtime, which allows cost savings when provisioning a lot of fine-grained application instances.

2. Container management platforms often provide additional features related to application property management, service registry and discovery, and secret management. It is possible to implement these functions on traditional IaaS, but you would either need to implement them with custom tools, like Hashicorp Consul and Vault, etcd, Eureka, or Zookeeper, or use proprietary cloud APIs. We will discuss this in more detail in the Microservices Platform chapter toward the end of the book.

3. Container management platforms make hybrid cloud deployments, and this makes it possible not only to migrate between cloud providers with the help of open source packaging and deployment tools but also makes it possible to migrate applications between private data centers with static servers and VMs and true IaaS.

4. To extend the previous point, container management systems can be used relatively cheaply to implement an interface between the infrastructure and application teams if the infrastructure team is using traditional data centers. Instead of going with VMWare and OpenStack, the team can set up Kubernetes and Mesos and achieve the same result.

Despite all of the benefits above, container management platforms are still less mature and generally weaker than traditional IaaS when it comes to basic management functions like scheduling, load balancing, networking, monitoring, and logging. You can achieve almost everything that you can do with container management systems by using VMs and IaaS cloud providers. First, remember that most of the existing personnel and tools are still more used to working with VMs than containers. Second, compare how much testing and budget goes into optimizing and testing the Amazon, Microsoft, or Google IaaS engines versus Kubernetes, Mesos, and Docker Swarm. Some of the container management platforms also still caution against hosting certain application types, like databases, in containers.

In any case, although we recommend the use of containers and container management systems for their benefits, we don't consider container management platforms as a fully fledged PaaS. They are just a more lightweight and more modern IaaS that oftentimes works on top of a traditional VM-based IaaS. Over time, they will become an integral part of any cloud provider's offering. To some extent, we already see this with GKE (Google Container Engine), Amazon ECS and EKS, and others. In this book, we don't separately cover function-as-a-service in detail, because we believe that the foundational principles discussed in the book would apply to this technology as well. Services like Amazon Lambda that allow the deployment and execution of pieces of business logic as functions are still niche offerings, but we believe that the trend towards serverless architectures is clear. It goes from treating VMs as servers, to VMs as executable binaries, to containers as tinier but still general purpose executable business, to pure business logic packaged and executed as individual deployable units. We will get back to the topic of deployable units and containers as deployable units later in the book.

If we return to the original notion of a PaaS, there is an assumption that it has many more components added, in addition to the base infrastructure and microservices platform (Figure 5.4.). These system components include databases, caches, message queues, and other functions available as a service. A PaaS often provides pre-configured middleware, like application and web servers, where the application packages are deployed. Essentially, the promise that a PaaS provides is that development teams can focus on writing application code and don't have to worry about all the other components that are needed for their application.

However, in our experience, this promise rarely holds and it comes with several major disadvantages. First of all, although PaaS is great for rapid development, when it comes to bigger applications, the default configurations of middleware, databases, and other components provided by the PaaS no longer satisfy the application requirements. Unfortunately, customizations of system components are limited and expensive when it comes to PaaS.

Second, although it is easy to use a public PaaS from service vendors like Heroku, it is difficult to build a private PaaS in house. All of the databases, caches, middleware, and other components need to satisfy various use cases from development teams and be available for provisioning and configuration via APIs. There are plenty of questions that need to be answered and implemented: for example, whether to go with multi-tenant component instances or with a single-tenant installation. The choice of multi-tenant dramatically complicates the management of these systems, whereas the use of single-tenant installations increases the number of potential custom configurations of component instances that need to be managed by a single team that doesn't know the specifics of the applications that use those components.

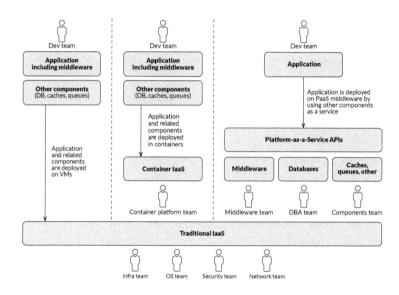

Figure 5.4. Traditional IaaS, container IaaS, and PaaS models.

Then, there are plenty of policies that we discussed at the infrastructure level. It is hard to implement them in a clean service-oriented way, even for infrastructure. Although an interface between applications and infrastructure is relatively clean and simple, an interface between applications, middleware, and databases is a lot more complex. Going up the stack increases the difficulty in providing a clean interface to satisfy all security, change management, and operations policies for all applications.

Finally, one of the reasons we have seen for why an enterprise may want to go the PaaS route is to utilize existing middleware and DBA teams. In our experience, this approach also doesn't work well, because these existing teams have been providing completely different services to what is required from PaaS. Historically, these teams just managed middleware and databases centrally with a human interface hidden by self-service portals. With PaaS, they will need to implement truly multi-tenant components that can be provisioned and configured via APIs. This is generally several orders of magnitude more difficult than what they were doing before. So, at best, these teams just become embedded in the respective application development teams and develop deployment packages with system components that are specific and customized to those applications.

We understand that the topic of PaaS usefulness may be extremely controversial. It is controversial for a good reason. We have observed many companies trying to adopt private PaaS technologies at scale. In the end, either the PaaS technologies were used at a small scale for a single team or only a limited set of capabilities were used. Specifically, scalable implementations that were successful didn't rely on the original PaaS promise of providing the full platform with system components, databases, and middleware. Instead, the platform-as-a-service was used more like a container-as-a-service microservices platform with some additional features, such as management and self-service UI. After all, there is a good reason why major PaaS technologies have started supporting containers as first-class citizens and have even replaced internal execution fabrics with open source CaaS tools.

We will not consider platform-as-a-service later in the book because there are very few successful large-scale implementations of true PaaS systems in enterprises. Although the situation may change in the future, we currently don't consider platform-as-a-service to be enterprise-ready in the way that infrastructure-as-a-service or container-as-a-service are. Often when a PaaS is used in an enterprise, it is the decision of a specific software development department or an application development team, which makes it essentially equal to the middleware and database components being a part of a business service. It doesn't matter at this point whether a database is an integral part of a service, such as part of a PaaS, or is a standalone component.

The situation with container management platforms is partially the same and partially different. Containers are not ready to replace traditional VM-based IaaS on the enterprise level, so oftentimes this choice is made by a software development team making a container platform as part of an application. On the other hand, CaaS is still just an IaaS with a difference in what counts as a deployable unit. This difference won't affect the primary discussion of this book, which is change management. For the majority of applications, it doesn't matter whether their deployable unit is a container or a VM. Most modern distributed applications, stateless and stateful, fall into one of these categories.

5.7. EXAMPLE: ECOMMERCE PLATFORM

Let's get back to the eCommerce platform architecture we outlined before and provide a conceptual deployment architecture diagram that we will use for our change management discussion. For the purposes of this example, we will assume that it is deployed to an infrastructure-as-a-service provided by Amazon.

Figure 5.5. represents the production infrastructure for the eCommerce platform. We assume that one cloud project is created to host all components of the platform. The network in the cloud is an extension of the enterprise-wide network in the private data center. From the outside, customers access the web UI via a content delivery network (CDN) and web application firewall (WAF). The cloud network is connected with the private data center via a VPN or Direct Connect. Inside the cloud, several subnets are configured:

1. The public subnet is used for external access from the Internet and contains only elastic load balancers provided by Amazon Cloud.

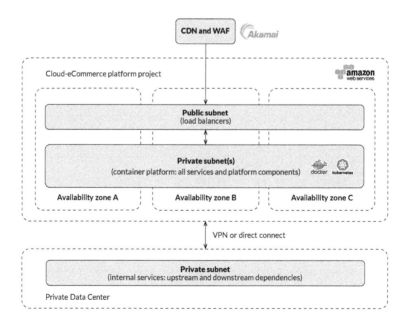

Figure 5.5. Production infrastructure for our eCommerce platform example.

2. The private subnet is used for a container platform implemented on Kubernetes. The container platform hosts all business services and system components, including business applications, databases, caches, messaging, and internal platform components like the service registry, feature flag management tool, monitoring, and logging.

3. Separate private subnets may be used for jump hosts for SSH access to VMs, as well as for proxies to let applications hosted in Kubernetes access services deployed in the private data center.

Details of the high availability, disaster recovery, networking, and security setup in the cloud environment for different applications is out of the scope of this book, and we'll just assume that the networks, subnets, and firewall rules are configured by the infrastructure team.

In addition to the production cloud account, we need to have separate accounts for test environments. This is usually the easiest way to isolate the non-production and production environments, access and networking wise (Figure 5.6.).

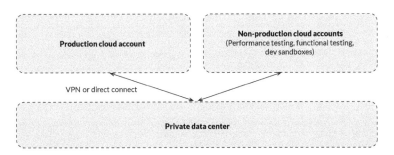

Figure 5.6. Separation of production and non-production environments with different cloud accounts.

From the access perspective, the following cloud APIs are exposed to the teams in the software development department:

1. Compute: create and destroy VMs, as well as edit VM meta-information like tags.

2. Storage: create and destroy storage buckets, and read and write into storage buckets.

3. Networking: create and destroy load balancers.

The software development department is often further split into development-focused and operations-focused teams. Although operations teams have access to both production and non-production accounts, development teams don't have access to production accounts.

The infrastructure team also provides a set of non-automated guidelines and conventions regarding tags that should be put on VMs to implement reporting and chargeback for financial accounting purposes. We also assume that, in addition to network-level pre-configured firewall rules, other application-specific rules are managed with VM tags. There are conventions and guidelines created by the networking and security teams that the development teams need to implement. The rules are enforced with audits.

In addition to exposing native VM-based cloud APIs, the Kubernetes-based container-as-a-service platform is implemented and managed on top of the cloud APIs. Separate container platform instances are created for different software development departments and for different environments within that department. The eCommerce department get their own dedicated instances of Kubernetes for production and different non-production environments. A separate team in the infrastructure department is responsible for creating and managing the Kubernetes clusters for the software development departments. Tagging, patching, and managing of VMs, storage, and networks for the Kubernetes cluster are tasks performed by the container platform team. Kubernetes clusters are single-tenant components and dedicated to the individual environments of each department, so security and access control are trivial. Specifically, the eCommerce release engineering and site reliability teams have full access to the clusters, and they set up more granular access and quotas for the respective service development teams.

6

ORGANIZATION

There have been many discussions in the industry about building the right organization and teams to facilitate DevOps culture and enable continuous delivery. An inefficient organization structure can have disastrous effects on the speed of delivery, quality, stability, and team motivation.

Due to differences in scale, the problems that surface in the enterprise environment are not visible in technology startups, so it is believed that technology startups have found a universal solution. Because of their lean size and emphasis on the pace of feature delivery, startups don't have the need, time, and resources to build architectural or organizational structures. So the organization ends up being as flat as possible. In extreme cases, this means an engineer is assumed to be:

- A full-stack developer, responsible for all tiers and components of a system, including web UI, mobile apps, backend services, and databases;

- Experienced in development, testing, and operations, oftentimes responsible for quality assurance, deploying code to production, and supporting it; and

- Self-motivated and self-managed to eliminate management layers.

This effectively means that an engineer plays all of the roles in a typical IT organization at the same time. This can work under some conditions, but we feel that accepting this situation as a norm for all companies is very dangerous. First of all, it is difficult to find people skilled and motivated to do every aspect of software delivery. Second, staffing a team with such people is a lot more expensive than the more traditional approach with division of labor. Even if very small technology companies can operate in such way, when they grow to tens or hundreds of people, they tend to implement some hierarchical structure and division of labor.

A flat organization with the engineers playing all roles is rarely justifiable for enterprises with hundreds or thousands of employees. Although optimization for a flat organizational chart may be intentional and carefully done in some companies, in most cases we have observed that lack of structure is a sign that the management have given up on doing their duty, have delegated all of their responsibilities to the engineers, and are left with no control over the situation. Leading such organizations becomes very difficult, the sense of a common goal often becomes lost, and efficiency and productivity suffer. Enforcement of company-wide security and change management policies in this situation is almost impossible.

However, flat organizations are not the most common problem. Most traditional enterprises are coming from the other extreme of very fine-grained division of labor and teams that are structured by functions instead of business domains and end-to-end goals. In this case, it is common to see tens or hundreds of very specialized roles, and each change that comes from the business requires effort from all of the roles. We have touched on the deficiencies of organizations with too high a division of labor structure before in this book.

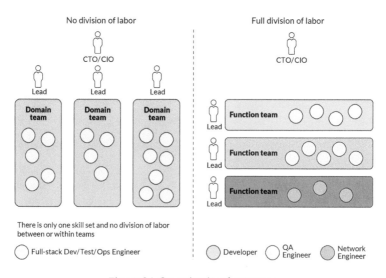

Figure 6.1. Organizational extremes.

To some extent, the extremes shown in Figure 6.1. resemble the classic organization structure options from project management. The model with no division of labor resembles a pure project organization, and the model with full division of labor resembles a purely functional organization. The similarity is not complete though because we also take into account the operations and division of labor inside teams, which is not in discussed in the scope of project management.

6.1. DIVISION OF LABOR

Although it is impossible to build a large organization without any division of labor, there are many different ways to implement division, decide where to draw the division lines, and decide how granular the division should be.

We have already discussed an IT and software development organization template for large enterprises, which nowadays lean toward a fine-grained division of labor. Although not perfect for DevOps and continuous delivery, this type of organization is relatively cheap. It is easier and less expensive to find one expert in web UI development, one expert in backend development, one expert in QA, and one support technician than four experts in all four areas. Unfortunately, a poorly structured organization with too fine-grained a division of labor causes many inefficiencies in other areas, and in the end, the price-to-value ratio may end up being very high while the actual cost efficiency suffers. In any case, we want to point out that division of labor is a powerful tool to control the cost and structure the team, so a new organization structure should embrace it and use it correctly. Too many times, we have observed unjustified efforts to build truly cross-functional teams, when the required combination of skills was rare and extremely expensive on the labor market. As new talent that would satisfy wide cross-functional requirements was hard to acquire and the existing team members couldn't become competent in all of the skills by management decree overnight, the project delivery speed and quality suffered immensely and the project often came to a complete halt. This, in turn, led to demotivation of team members, loss of trust from top management, and ultimately, failure and reversal of the whole transformation effort.

Although cost reduction and the availability of specialized resources on the market is a clear benefit of division of labor, there is one subtle but major disadvantage. If the division is very granular, a person who is responsible for a very specialized function will end up being too far away from the end result of his or her labor. That person will not be able to associate their task with the bigger goal, which typically leads to several issues:

1. Low motivation. When a person doesn't see the end result, the task loses its meaning, and human beings generally don't like to perform meaningless work.

2. Lack of context. It is oftentimes hard to fully formalize and describe the task given to a person in complete detail. Something will be missing, and without the broader context of the problem that the team or organization tries to solve, either quality will be low or work will need to be redone.

3. More handoffs between specialists (production stages) and associated delays.

All of these issues ultimately lead to low quality, low efficiency, and high management overheads.

In order to build an efficient organization, a middle ground should be found between the division of labor being too fine grained or too coarse grained. We will discuss this next.

6.2. SERVICE-ORIENTED ORGANIZATION

There is no question that in order to scale up, a manageable organization needs to have some hierarchical structure. One of the approaches that, in our experience, works for enterprises is to build a service-oriented organization (see Jeff Bezos' mandate [25]). In the cleanest form, this type of organization demands that every team needs to provide a service via an API and prohibits any day-to-day communication except through that service and the API. The services can be internal or customer-facing.

Making infrastructure available as a service is one example of how to facilitate a service-oriented organization. When it comes to software development departments, this means that teams can be naturally aligned with the services that they develop, which effectively leads to the implementation of inverse Conway's Law [26, 27].

Enterprise IT and software organizations can be very large, consisting of thousands of people, so this service-oriented model can have a nested hierarchy. First of all, let's consider the top-level view in Figure 6.2.

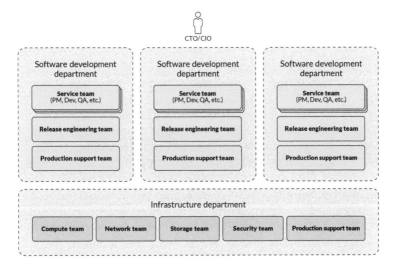

Figure 6.2. Top-level view of a nested hierarchy in a service-oriented organization.

Each department provides a number of software services via an API. On this level, we try to follow the service-oriented organization principle as closely as possible. The infrastructure department and related services have been discussed in depth before, but the software development departments also deliver a set of services within a large business domain. For an eCommerce example:

- One of the software development departments manages the eCommerce platform discussed in this book.

- Another department can be responsible for fulfillment, providing order management, inventory management, and logistics services via an API to the eCommerce platform.

- Yet another department can be a business intelligence (BI) department, providing an analytical data platform (data lake) and BI.

- Other departments may include various business systems, like ERP.

The same concept applies to the infrastructure department, which was proven in the best way by Amazon when they first implemented cloud-like functionality for internal use and then exposed it as a service. However, traditional infrastructure departments in enterprises may not be ready to provide a service as-is. As we have discussed before, some functions, like middleware and DBA, are better aligned with applications. In this case, we recommend making them a part of the software development departments and specific applications teams. The rest of the infrastructure department can provide the regular IaaS services discussed above. From a change management perspective, this team will implement changes that concern only infrastructure. For example, if an infrastructure department is still managing private data centers, the migration to a new datacenter or upgrading of a data center is easier to do with a single department. Another example is networking: because services implemented by software development departments may need to communicate over a private and secure network, there is typically an enterprise-wide network that is more easily managed centrally by a single team.

Let's look at the structure of each software development department (Figure 6.3.). Internally, it should follow the same principles of service-oriented organization.

The software development department is split into a number of teams, who are responsible for implementing business services. By design, such services provide their functionality via well-designed APIs and strong functional and non-functional contracts, which are enforced by tests. Each service development team is led by a manager or director-level person and comprises a number of familiar roles, like product manager, project manager, architect, development lead, QA lead, developers, functional QAs, non-functional QAs, and deployment engineers. The involvement of information security (InfoSec) personnel in service teams is important to enable development security operations, although it is rare for each team to have its own dedicated security specialists. Most probably, it will be a part-time allocation from an enterprise-level security team. The specific division of labor within the team is up to the team lead, on the basis that cost constraints can be met and a certain efficiency can be achieved. In practice, several roles may be played by the same person, as long as this doesn't violate segregation of duty policies [28].

Although release engineering and production support teams can be embedded within software development teams, we recommend keeping them separate:

- Release engineering should be separate because it enforces company-wide and department-wide change management policies. We will talk about how a release engineering team communicates with other teams via a service-like interface later in the book. It is important to note that the release engineering team we discuss here doesn't participate in the day-to-day manual approvals of changes to production. Instead, it designs and enforces the process that the service team need to follow and implement.

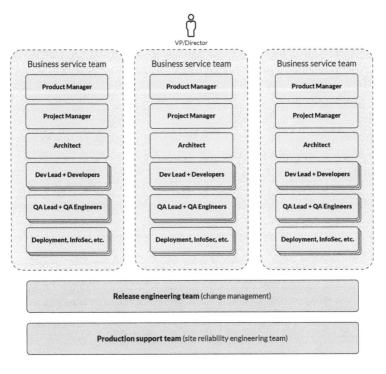

Figure 6.3. Service-oriented organization of a software development department.

- Production support should be separate for division of labor and cost efficiency purposes. This team typically needs different skills, use its own set of tools, including monitoring, logging, and alerting, and is required to work with different policies. Over a period of time, the production support team evolves into a site reliability engineering team. L3 support often stays within service teams, because it requires good understanding of service functionality and deployment specifics.

There are additional roles and teams that may be required at the top level and department levels. Such roles are required for establishing and enforcing compliance with external and internal enterprise-wide policies and guidelines. Although such teams may be harder to fit into service-oriented organizations, it should be done when possible. The same approach to any compliance policy enforcement should be used as that in cases of regular test automation.

By designing the organization in the way shown above, we can achieve cost efficiency with a certain division of labor, but at the same time, each team will be more empowered and will see the end result of their work, because each team provides a service that is consumed either by internal or external teams. By implementing a well-defined hierarchical structure, the whole organization becomes manageable. Each department and team can have their own metrics and KPIs related to quality, stability, efficiency, cost, and speed of reacting to change. At the same time, the organizational structure is not dramatically different to how existing IT organizations are structured, so it won't require a complete redesign that is expensive and risky.

6.3. CULTURE

Culture is oftentimes presented as a silver bullet in solving the problem of DevOps, continuous delivery, and otherwise efficient organization. Unfortunately, culture is difficult to define with precision. In our experience, although culture is important, we consider it more of a symptom that something is led and managed well. It is difficult to improve culture directly and to just set up the right culture by issuing an executive order.

Rather, culture is a response to the environment in which people are put. This environment is oftentimes characterized by the available tools and resources, external and internal constraints, leadership, and goals. This is true both for society as a whole and for a company. So, a culture can be created by creating the right environment and maintaining it for a period of time that is long enough for people's behavior to change. This means first fulfilling the right prerequisites, giving teams the tools they need, empowering them with enough information and context, measuring how they do with the right metrics, and guiding them in the right direction with KPIs. In a sense, the creation of an organizational culture is like the forming of good personal habits.

In our experience, if the leadership tries to set the right culture in a very explicit way without creating the right environment, it can trigger an adverse reaction. In the best case, instead of truly changing the culture, the leadership will create a vision of the culture, in which the external aspects of the culture will be simulated, like a facade, without the internal mechanics of the team's behavior being changed. In the worst case, we have observed how, in some companies, team members will "hack" the leadership's understanding of culture and dedicate all effort to simulation, thereby lowering organizational efficiency even more and completely derailing the company from making any actual progress.

In order to be manageable, each department and team needs to have a number of metrics and KPIs, which allow efficiency to be measured and the organization to be steered in the right direction. We will discuss specific examples of metrics and KPIs in a separate chapter later in the book. They can be formal or informal, but they have to exist. They have to be measurable, and there should be control points to enforce compliance in case some teams start going in the wrong direction. To some extent, the right culture can be considered as a set of informal metrics and KPIs that all of the people in the organization understand consciously or subconsciously. But even culture should be monitorable and enforceable at all levels of an organization. Otherwise, even if there is a great culture set in the beginning, it will deteriorate as the company grows or the management changes.

The focus of this book is to establish the right prerequisites in the form of organizational structure, architecture, policies, processes, metrics, and KPIs. We don't talk about culture explicitly to avoid a "cargo cult." We believe that, after establishing the right environment, positive changes in the culture will follow.

6.4. DEVOPS TEAM

In the new organization, the role of the release engineering team changes. Instead of being manual orchestrators of the change management process, they implement the service that automates it. Often, when release engineers cannot implement such automation, companies will create new teams called DevOps. Although we defined DevOps as an organizational culture earlier in the book, we can't deny the fact that many companies are using this term to denote a team or a role. In this sense, the task of a DevOps team is to provide automation in two areas where the existing teams do not have the skills to do it: change management / CICD pipeline automation, and environment provisioning and deployment automation. Let's review these two areas separately.

Deployment automation is more aligned with the application development teams. If developers can't implement automated deployment due to skill gaps, a special role for deployment engineers can be created, and those engineers should be embedded in the development teams. When multiple teams in the company are relying on the same system components, like databases and caches, development teams can reuse the deployment packages of common system components. Such reuse can be implemented and facilitated like an "internal open source model." The application teams, who are the first adopters of a new technology, can create an initial distribution. After application, teams can take base packages, create their own forks, add desired customizations, and contribute changes as needed. Contribution to such reusable components should be measured and rewarded on a company level. It typically doesn't require significant facilitation, because the open source model is already familiar to the majority of engineers in the industry.

Change management, or CICD pipeline, automation is naturally a task for the release engineering team. If that team currently doesn't have the skills, it should be enhanced with the required skills. It may be lucrative to create a new DevOps team parallel to the RE team, and in our experience, this may work in the short term to bootstrap transformation, but it proves to be inefficient in the long run due to misaligned priorities. A separate team doesn't prevent a gap between the "modern development" represented by the DevOps team and the "traditional operations" represented by the release engineering team. So, even if a separate DevOps team is created to help with establishing continuous delivery initially, it should ultimately be merged with or become the release engineering team.

6.5. EXAMPLE: ECOMMERCE PLATFORM

Let's continue following the example with the eCommerce platform. For this example, we will show one top-level software development department and one top-level infrastructure department in detail (Figure 6.4.).

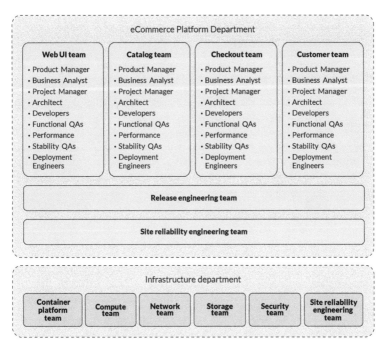

Figure 6.4. Organization of a software development department and an infrastructure department for our eCommerce platform example.

The infrastructure department provides infrastructure-as-a-service, as discussed previously in the book. Infrastructure has its own SRE team that operates and supports infrastructure services. The infrastructure SRE team is also responsible for monitoring and logging, as well as incident and problem management tools. The scope of these tools though is limited to infrastructure services, and the eCommerce platform team is free to choose its own tools if needed. The reusability of such tools should be encouraged, but not dictated, at the company level.

The infrastructure team also provides a container platform service, which represents a container-based IaaS and allows deployment and running of containers instead of VMs on the infrastructure. The eCommerce platform department has four service teams:

1. Web UI team. Responsible for the responsive adaptive web UI service, including a Node.js runtime environment.

2. Catalog team. Responsible for the catalog and catalog import applications, including middleware, catalog database, and catalog feed components.

3. Checkout team. Responsible for the checkout service, necessary middleware, and cart and order databases.

4. Customer team. Responsible for the customer service, necessary middleware, and customer database.

All service teams have their own product managers, business and systems analysts, project managers, architects, development leads, QA leads, developers, functional QAs, performance QAs, stability QAs, and deployment automation engineers. The teams are fully responsible for developing, testing, and automating the deployment and configuration of their services and respective components. Within some teams, several roles can be played by the same person if these roles correspond with their skills, although most of the time people will have specializations. In addition to full-time roles, security architects and engineers are assigned to consult with service teams on a part-time basis. Any reuse of components from the deployment and configuration perspective is encouraged, but ultimately, service-level deployment teams are responsible for the configuration of their components in the best way for their services.

The release engineering team is responsible for implementing the continuous delivery process and pipeline according to the change management policies. The next chapter will be dedicated to the details of this implementation.

The site reliability team is responsible for production operations and support, as well as for the setup of monitoring, logging, alerting, and incident and problem management tools, and their configuration. In addition, site reliability engineering is responsible for other platform-level shared components, such as the service registry, secret storage, and messaging infrastructure.

7

CHANGE MANAGEMENT

The primary idea behind implementing continuous delivery is to stick with existing change management policies but to take a fresh look at the existing processes, procedures, and tools that implement them. Some old procedures can be discarded and new ones can be implemented to take into account the microservices architecture, cloud, new service-oriented organization structure, and end-to-end automation.

With continuous delivery, the release engineering team, instead of manually orchestrating releases, will need to create and formalize a fully automated process and define formal service-oriented inputs for the process. After doing that, the team will start providing the change management process as a service and will consume inputs from other teams as a service as well. As a disclaimer, we can say that this doesn't necessarily mean that all of these services will have to be exposed via a REST API as a hard rule. Occasional manual actions may still be required, but they should satisfy strict contracts and SLAs and should otherwise fulfill the requirements to be a service when possible.

7.1. INVERSION OF CONTROL

As we discussed earlier in the book, existing change management policies can be summarized as follows:

1. The initial change should be requested by an authorized person.

2. Before implementation of the change in production, it should be prepared and signed off by a number of designated people

in the organization, including the development lead, QA lead, security lead, and operations lead.

3. After implementation of the change in production, it should be verified by the test lead and operations lead.

4. An audit log should be available for any change.

In a traditional implementation of the process, the release engineering team is responsible for collecting all necessary approvals and preparing the change before it is handed to the operations team to be implemented in production. This is an inefficient and poorly scalable process, because it involves significant communication between various teams.

Figure 7.1. *A typical step in a traditional change management process.*

For example, a typical step in an existing change management process is when a release engineer needs to receive a sign off from the QA lead that the change is good. Even if communication between the two roles is automated in email or JIRA, the process and the step itself require manual involvement from both roles (Figure 7.1.).

The step above is executed for every run of the change management process, which is done for every change. In the spirit of service-oriented organization and in order to make the process more efficient, this flow should be reversed, as shown in Figure 7.2.

In this case, both teams implement and provide a service. The service that implements the change management process uses the QA service to execute QA policies that the QA team have defined for a certain type of change. Technically, the QA service in this case implements QA procedures (as a process), not policies. However, in this and other cases, it is easier to say that a service implements a policy.

At the next level of detail, QA policies can include test data, automated tests, and test analysis and acceptance criteria (Figure 7.3.). The service to deliver this via the API can be as simple as Git or as complex as a test farm that would execute the required tests against a specified environment. Implementation of release engineering policies can be as simple as a series of Jenkins jobs, one of which will implement the step of testing a change.

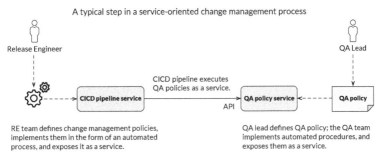

Figure 7.2. A typical step in a service-oriented change management process.

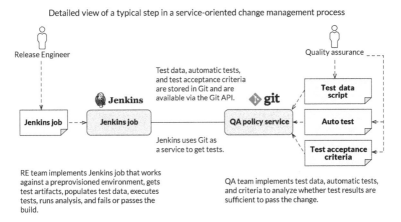

Figure 7.3. More detailed view of a typical step in a service-oriented change management process.

At this point, the picture starts to look more like the traditional continuous integration that is described in many books. The major difference is the approach, together with the fact that QA policy service needs to be fully self-sufficient. Too often, in our experience, the only part that is automated is the actual test, with the test data management and test acceptance criteria remaining manual and requiring manual intervention. We will get back to environment provisioning and deployment, as well as the importance of automation of the test data, test analysis, and acceptance criteria, when we get to the specific end-to-end process of handling changes.

The approach we described, with a testing step for the process, should be applied to all other sign offs and approvals that the release engineering team is collecting manually.

There is one principal conceptual aspect that needs to be taken into account and internalized by all teams participating in change management. In the old process, they were asked to provide manual approval, and then the workflow, with manual steps of requests and approvals, was automated with various business process automation tools. There was significant leeway in how the teams could provide the approvals: they could automate the approval process internally, but most often they still required a human review of the results of that automation before providing a sign off.

In the new process, instead of being asked for approval every time, the teams expose a service and an API implements the approval decision making and allows other tools to get such approval at any time it is required. This powerful technique is called inversion of control and it is being used very frequently in software engineering. We are now applying it to the organization and change management process.

Implementation of this technique may present some challenges. Specifically, in the case of test automation, the tests themselves are oftentimes automated. The challenge is that the confidence in the tests may be low, they are not automated for 100%, and most importantly, test analysis is still manual. However, we would argue that there are technologies and skills in the industry to close these gaps. We'll discuss the test automation example later in the book when we get to the detailed review of service-level and integration testing.

7.2. PROCESS TEMPLATE

With the inversion of control, the new process represents a workflow for collecting all approvals for every change, plus a number of auxiliary steps to build artifacts, provision environments, deploy applications, etc. (Figure 7.4.).

Certain steps in the process are not automated by design. For example, a request for a change is a naturally manual process, because it originates from a business stakeholder. Development of the change cannot be automated. But most steps after the change is developed can be streamlined, so that almost no human intervention is required. Another step that may be left as manual is the triggering of the deployment of the change to production. The actual deployment process and later verification process can be automated, but the trigger still remains in the hands of the production operations team. This is oftentimes the time to do a final review to ensure that the process defined and implemented by the release engineering team works correctly, all approvals are in place, and the change is otherwise ready. Inversion of control in this step is not the top priority, because it is not a source of major inefficiencies.

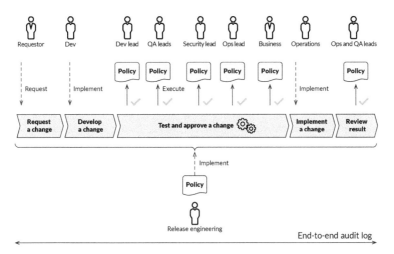

Figure 7.4. Process template with inversion of control.

At the center of the new process is a completely automated continuous delivery pipeline that is implemented by release engineering according to change management policies. There are several inputs for this pipeline, as defined in Table 7.1.

Some of the policies may be harder to automate in the beginning. One of the policies that may continue being manual is the final business approval. Such approval is typically done by executive management or product management after user acceptance testing (UAT). It is highly unlikely that either executive management or product management will automate this policy. However, because they are the original requesters of the change, it is also unlikely that there will be any concerns or major inefficiencies from their side when it comes to change management.

We'll discuss all steps in the pipeline, including policy inputs and automation opportunities, in detail later in the book.

Table 7.1. Inputs to an automated continuous delivery pipeline.

Team	Equivalent to approval	Inputs to CICD pipeline

Development	Dev lead	* Build scripts
		* Deployment scripts
		* Unit tests
		* Unit tests analysis script and policy:
		    Minimum code coverage
		    Acceptable percentage of failed tests by test type
		    Acceptance test failures
		* Static code validation tools
		* Static code validation rules and configuration
		* Acceptance failures of static code validation

Functional QA	QA lead	* Test data generation scripts
		* Automated tests:
		    Environment validation tests
		    Smoke tests
		    Full regression tests
		* Test analysis script and policy:
		    Minimum code coverage
		    Acceptable percentage of failed tests by test type
		    Acceptance test failures

Performance QA	Performance QA lead	* Test data generation scripts
		* Automated tests
		* Test analysis script and policy:
		    Acceptable percentage of failed tests by test type
		    Acceptance test failures

Security	Security lead	* Test data scripts * Automated tests * Test analysis script and policy: Acceptable percentage of failed tests by test type Acceptance test failures
Operations	Operations lead	* Validation policy that all approvals are collected * Automated stability tests * Test analysis script and policy: Acceptable percentage of failed tests by test type Acceptance test failures * Release notes review
Business	Product manager	* Release notes review * User acceptance tests

7.3. TYPES OF CHANGE

The template of the process may vary depending on the type of change. At the beginning of the book we mentioned several types of change that are subject to change management processes: application code, application runtime properties, application business properties (feature flags), middleware, operating systems, infrastructure (compute & storage), and networking.

With the new architecture and organization structure, we will group them into different types:

1. Application code, certain application configuration, and middleware (mostly application server) changes are considered to be business service changes, because both code and middleware are being implemented by development teams. These changes are most complex and risky from the business standpoint and require all of the steps in the process template. The good thing is that these changes are environmentally agnostic and can easily be tested.

2. Application runtime (endpoint) and business configuration (feature flag) changes are either environment specific and hard to test in general or they are tested in various combinations as part of the type 1 changes. We will argue that these changes may not be even considered changes at all and, at the least, they require special treatment and a very lightweight process.

3. Compute, storage, operating system, and networking changes are considered to be infrastructure changes. These changes may come from different angles: they can be either application related, like scaling in and out, or infrastructure related, like applying new security patches to operating systems.

As mentioned before, we do not consider changes in tests, test data, or production data as changes that need to be managed. The case of production data is obvious, and when it comes to tests and test data, they don't satisfy the criteria to be changes: they are not deployed to production and do not affect the production system.

We will also consider changes to business configuration of scalability, failover, and self-healing events to be very special types of change or not even changes at all. We will discuss that in detail later in the book.

7.4. SERVICE CHANGES

We will start with the end-to-end change management process for business service changes, from the request for a change by the business until the deployment to production. The scope of this discussion is not limited to business applications but rather can be applied to change management for all components of services that include business applications, caches, message queues, databases, and so on.

When we discussed architecture and organizational structure earlier in the book, we used the following hierarchy:

1. Enterprise.
2. Business domain.
3. Service.
4. Component: system (cache, database, message queue) or business (custom application code).

In this hierarchy, most of the changes requested by a business are limited to a single service. We assume that large business features that affect several services are broken down into smaller features that affect one service each. Later in this chapter, we will discuss cases when individual service-level changes can be deployed in production independently but in parallel and when there is a strong dependency between changes in different services.

Although from a business perspective, a service is an atomic and indivisible entity, from a technical perspective, a service is composite. Each service consists of a number of components, typically including an application with custom code implementing business functionality and several system components, such as databases, caches, and message queues. Each service-level change affects one or more of these components on a physical level. As business application components change more frequently than system components, it is beneficial to link changes to components and implement some pieces of change management on the component level. For example, if a change affects only the business application component, this changed component may be built and deployed to production on its own, with the other components of the service left unchanged. This will reduce the number of entities affected by each change and will help to reduce the risk profile and increase the efficiency of the change management. However, we should keep in mind that component-level changes are a technical optimization and the actual change management from the business perspective is still performed at the service level.

The granularity of changes affecting pieces smaller than independently deployable components is not considered. Although some changes may affect only one line of code, one class, one source code file, or one library, there is typically no well-defined testable contract between these entities. Dependency management, configuration management, and individual deployment of such fine-grained changes bring more challenges than benefits.

Changes are linked to services, so change management implementation should take into account the affinity of a business change and a service. This affinity should be maintained throughout all stages of the process. Source code repositories, deployment packages, and tests need to be designed to enable service-level granularity of changes.

7.4.1. Business Requirements

The lifecycle of a change starts with a business requirement. Some business requirements may correspond to new features and some may be related to improving certain aspects of an application, including performance, stability, security, or maintainability.

According to change management policies, only authorized personnel can request a change. In software development, the product manager role is responsible for initiating changes or approving changes requested by others (Figure 7.5.). There are different methodologies, including the familiar Scrum and Kanban, to manage the wish list of changes and prioritize the important ones for execution. This aspect is covered well in many project management books and we won't go into details here.

In the initial stages, requirements for software development can be managed in documents or with specialized tools. However, when the time comes to actually start implementation of the changes, requirements are created and managed in project management tools, like Jira, Mingle, and many others. We may use Jira as an example in this book, but everything we discuss also applies to any other tool.

If a change is created in a project management tool like Jira by creating a ticket (story or task) by a person in a role other than product manager, the product manager still needs to approve the change before it will be taken for implementation. In the case of Jira, this can be done by either allowing only the product manager to transition the ticket from "created" to the "approved for implementation" stage or by requiring the product manager to leave a comment; in the case of Scrum, the product manager can be required to transition tickets from product backlog to sprint backlog. Although implementation may vary, there should be an audit trail that a requirement was approved.

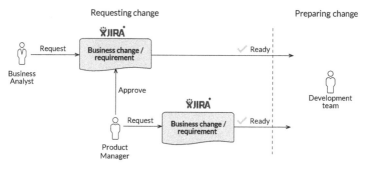

* In this and other pictures later in the chapter, the technologies given are not the only options, just illustrative examples.

Figure 7.5. Role of the product manager in initiating and approving changes.

After being requested, an application change will go through a complex process of development, packaging, and testing. During this process, a single business change can be packaged together with other business changes and that combined package will actually be deployed to production. With all of these transformations, it is important not to lose track between the deployed package and the original change requested by the business and to continue speaking the same language across the board. After all, the business is less interested in the specifics of packaging and more interested in the change as it was requested and tracked in Jira or another tool.

We should also point out that, in this case, we are talking about requirements to services rather than business application components. This means that a requirement may affect multiple components in a service.

7.4.2. Source Code

As we mentioned before, a change may affect a business service, which may consist of multiple components including business applications code, databases, caches, queues, etc. At the same time, the application development team is responsible for everything in the application above the infrastructure layer (Figure 7.6.).

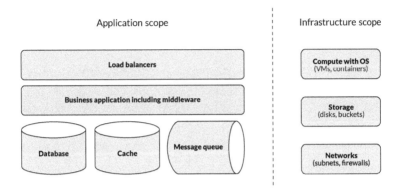

Figure 7.6. Scope of applications and infrastructure.

This means that the application source code includes:

1. Actual business application source code implemented in one or more programming languages (Java, JavaScript, C, Python, PHP, etc.).

2. Build scripts for business application source code to produce deployable artifacts.

3. Business application configurations, including the template for configuration and specific values for environmentally independent items. A common implementation approach is to define such configurations as property files with property names only for environmentally dependent variables and property names with values for environmentally independent configurations.

4. Provisioning and deployment scripts, including those for the deployment of all business application components and the relevant middleware. We use the term "script" generically for simplicity and familiarity, although we understand the declarative deployment configuration manifests from tools like Puppet and Terraform are not strictly scripts. We try to avoid abusing the term "configuration management," because it is too wide and can mean too many different things depending on context.

5. Provisioning and deployment scripts for load balancers, disks, storage buckets, and other infrastructural components of a service.

6. Provisioning and deployment scripts, as well as configurations, for system components that are dedicated to the service, like databases, caches, and queues.

7. Shared libraries that applications in this service or across different services may use.

8. Function, performance, stability, and other applicable tests for the service, including test data generation scripts.

Essentially, the service source code should include everything that is required to build, deploy, and run every component of a service. From a source code management perspective, either all of the parts of the source code can be put into one repository or each of them can be in its own repository or even several repositories.

There are several drivers behind the decision of how to structure source code repositories. From one perspective, everything from the list above is a part of a single business service, so all of those parts are relatively tightly coupled from a change management perspective. This means that we at least need to have a mechanism to manage dependencies and versions of these parts together. Therefore, repositories that are too fine grained would require excessive dependency management.

Coarse-grained repositories help with dependency management, because all parts are versioned together. At this point, we assume that a single repository produces a single deployment package in the course of a build process. Such a deployment package is a self-sufficient package that can be deployed on any environment without any additional information, taking only an environment-specific configuration as an input. We assume that if one component depends on another component in the same repository, this dependency will be resolved during the build process, meaning that the components will be built together, will be packaged together, and will ultimately share the same version.

Repositories that are too coarse grained, however, may lead to other issues. For example, different components of a service may be deployed to production separately. There is no need to deploy a new version of a database to production if it hasn't changed from the last version. Some parts of the source code (tests and test data) do not even come under the change management process. So, if a change is made in a coarse-grained repository, we need to ensure that only affected components are built and new versions of only affected deployment packages are created and later deployed to production. This functionality is supported by many build and deployment tools, but in our experience, it is harder to implement it properly, especially when multiple components of a service are built with different build tools.

7.4.2.1. Repository Structure

A balanced approach typically works better when a separate repository is created for different deployable units. Each deployable unit should roughly correspond to a running instance of a component running in production (Figure 7.7.). However, a single repository should contain everything that is required to build a self-contained package that can later be deployed to any environment. The exceptions to this rule are shared libraries, which are not deployable units by themselves.

Note that infrastructure components, such as the load balancers, domain name system (DNS), and storage buckets, are not considered to be separate deployable units. Instead, they are considered to be integral parts of their respective components.

With this approach, it is possible to isolate changes to business applications that change often and those to system components that change rarely and are typically riskier to change. However, this approach will require dependency management between the components of a service. This can be implemented with build-time dependency management by choosing a primary component, which is typically a business application, and referencing system component versions from it. Other approaches include the creation of a separate repository with component versions for a service or the creation of a global version repository. We'll return to this topic later in the book.

An alternative approach would be to put all components in the same source code repository. In this case, the build and deployment tools will need to be configured to create new versions of deployable units for which the source code is affected by the change and to deploy only those deployable units later in environments. This is certainly feasible and most of the heavy lifting can be done by the tools. It will require more careful reasoning and audit logs about what was

changed in each specific deployment and what didn't change, in order to avoid potential situations where a critical system component
goes down for an upgrade during a deployment that wasn't sup-
posed to affect it. Because of these challenges, we generally don't
recommend working with such coarse-grained repositories.

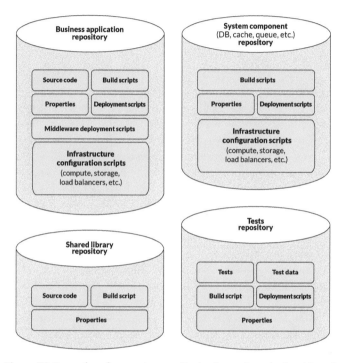

Figure 7.7. Examples of separate repositories for various deployable units.

Going forward, we will assume that the former approach to source
code repository management is used, with one repository roughly
correlated to one component / deployable unit, but the concepts
described in the rest of the book apply to the latter approach as well.

7.4.2.2. Build-Time Dependency Management

In addition to the runtime dependency management that exists
between services and components within services, there is build-
time dependency management. We have eliminated most of the

explicit build-time dependency management by combining different parts of a deployable component into one source code repository. Sub-components within that repository don't need to reference specific versions of each other. This is easy to implement in any build system and doesn't require a lot of maintenance.

However, there is one case where explicit build-time dependency management is required: between components representing deployable units and libraries. Shared libraries may exist both for business applications and for system components. For example, if we develop a business application in Java, there may be a shared library that is used between business services in the company to facilitate reuse. If a company uses a Cassandra database in many services, there may be a repository with reusable deployment and configuration scripts that follow company standards and enable easier onboarding for new business services.

Because libraries are not deployable units in themselves, they need to be embedded into deployable components during the build process. In this case, specific versions of libraries' artifacts need to be referenced and enforced during the build process. Specifying wildcard versions of such libraries like "latest" or "snapshot" will lead to non-reproducible builds of deployable components. The code of a deployable component may not have changed, but because the latest version of a library has changed, a new build in a deployable component may produce a new artifact with changed functionality. Most build tools support specifying exact versions of dependencies, but the enforcement may require a peer-review process and, in some cases, custom enhancements in the tooling.

7.4.2.3. Branching

Working with source code repositories requires a certain branching strategy. The most popular strategies with Git are Git Flow [29] and GitHub Flow [30]. They vary in how many types of branches they require, with Git Flow providing a more robust and controlled environment at the cost of managing more branches. GitHub Flow is leaner and allows faster delivery cycles.

In our experience, branching is too often abused to manage individual business changes separately from one another and isolate development on different changes between developers and development teams. With a microservices architecture and a single repository per component, library, or test, excessive branching brings more harm than good. Isolation is already inherent in the architecture and implemented at the repository structure level. Additional branches come with cost or merging; this delays integration issues, increases the feedback loop to find defects, and ultimately slows down the change management process.

Another approach to address the isolation of individual changes is feature flags [31]. In this case, the development team may be working on multiple features in the same codebase, but each feature is protected with a feature flag. With feature flag protection, incomplete functionality can still be deployed to production, but it will be disabled. Feature flags come with their own cost of increasing the complexity of the code and making testing of the application harder because all reasonable combinations of feature flags should be covered with tests. Feature flags also have their own lifecycle, and after the functionality is working stably in production, they should be retired via refactoring.

There is still a debate in the industry, with both branches and feature flags having their own advantages and disadvantages, but in our experience, more and more companies are leaning toward a single-trunk approach with feature flags, thereby sacrificing the additional cost of development and maintenance in favor of the increased speed of the change management process and the ability to deal with integration issues as early as possible. Ultimately, the decision between branching or feature flags is an open choice that any team can make on their own. Sometimes teams that are competent with branching and merging in a specific code versioning system would prefer that approach. However, practice shows that, for teams not familiar with branching, Git Flow, or GitHub Flow, feature flags are an easier way to achieve a fast and efficient change management process, so our general recommendation is to use the single-trunk approach. We'll assume this approach is used going forward in the book.

With this approach (Figure 7.8.), the development follows the single-trunk (development or master) branch. All commits go to the trunk after pre-commit validation and code review. When a commit becomes a release candidate after the CICD pipeline successfully finishes, it is marked with a tag. Alternatively, a branch is created instead of placing a tag. If a hotfix needs to be deployed to production, but the current trunk happens to be broken, the fix can be committed to the trunk and cherry picked into the branch corresponding to the version currently deployed in production. Alternatively, a hotfix can be committed directly to the branch with the production version, but in this case, a back-porting merge back to trunk will need to happen. Because a merge can lead to conflicts, cherry picking is preferred. When a new commit to trunk becomes a release candidate, tags or branches with previous releases still remain.

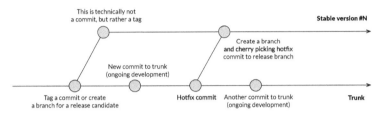

Figure 7.8. Single-trunk approach to development.

7.4.2.4. Example: eCommerce Platform

Let's consider an example for our eCommerce platform. Figure 7.9. shows the source code repository layout for the components we discussed in the Architecture chapter.

We omit other system components like the service registry and business configuration repository for simplicity. However, they follow the same rules as other system components and services like messaging. The repositories in Figure 7.9. correspond to those described in the repository structure chapter above (Table 7.2.).

In the case of build-time dependencies, referencing of specific versions is required. For example, the catalog application needs to reference the backend library via a specific version. If we assume that both of them are implemented in Java and Maven is used for building code, we can assume that this build-time dependency will be managed by pom.xml. Special enforcement of the use of specific versions may be required during the build process.

We create test repositories only for service-level tests. We do not explicitly and exhaustively test individual components, like MongoDB or Cassandra, and we rely on basic health checks that we inserted into the deployment scripts. This saves the labor of test engineers. From one perspective, these are open source components that have already been tested by the community. From another perspective, they will be exhaustively tested by service-level tests when deployed together with the business applications that use them.

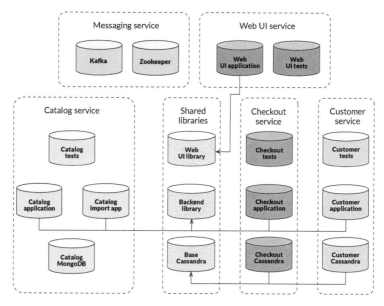

Figure 7.9. Source code repository layout of the components for our eCommerce platform example.

Let's look into the base Cassandra repository in more detail. According to the architecture, two services from the eCommerce platform use Cassandra as a database: checkout and customer services. We want to facilitate reuse, so we create a library repository with a base deployment Cassandra cluster. Building of this repository will not produce deployable units, but it will produce artifacts that can be referenced and extended for specific services. We will discuss deployable units in the sections about build and deployment later in this chapter. Specific versions of Cassandra will override the properties and configuration of the base Cassandra deployment for the needs of specific services. If the differences between Cassandra deployments across services are too big, reuse can be achieved by forking the base Cassandra repository. However, this will make it very hard to port new features and bug fixes from the base Cassandra deployment.

Table 7.2. Repository types and their contents for our eCommerce platform example.

Type	Repositories	What is inside
Business application repository	Web UI application Catalog application Catalog import app Checkout application Customer application	Source code Build scripts Properties Deployment scripts Middleware deployment scripts Infrastructure configuration scripts
System component repository	Kafka Zookeeper Catalog MongoDB Checkout Cassandra Customer Cassandra	Build scripts Properties Deployment scripts Infrastructure configuration scripts
Shared library repository	Web UI library Backend library Base Cassandra	Source code Properties Build script
Test repository	Web UI tests Catalog tests Checkout tests Customer tests	Tests Test data Properties Build script Deployment script

7.4.3. Code Review

The code review process ensures that the change requests are peer reviewed and approved by designated senior developers before they go to production. Code review is typically implemented very early in the continuous delivery pipeline. The peer review requirements for changes are one of the key change management policies. They are defined and enforced by release engineers as a part of the pre-commit pipeline to ensure that new changes in the code are easily identifiable and that the unreviewed changes cannot physically get to the main development branch.

7.4.3.1. Pre-commit Pipeline

In addition to the code review, the pre-commit pipeline also contains build and test phases (Figure 7.10.). This is not strictly required by the policies, but it is an optimization mechanism that helps code reviewers to block bad changes and increase the stability of the main development branch. It also helps the developers who initially submitted the change to get feedback about the quality of code early in the process. At the same time, it is important to ensure that the test phases of the pre-commit pipeline are fast enough not to become a blockage and to finish before the code review.

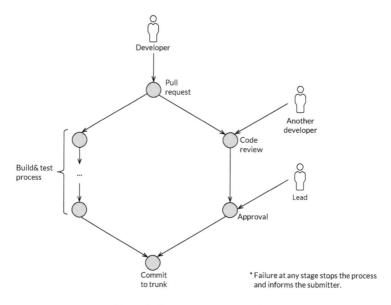

Figure 7.10. A pre-commit pipeline.

We will review the build and test process in more detail in later sections. Typically, the pre-commit pipeline includes only build, static code validation, and unit testing. Functional and performance testing is often omitted to avoid long wait times early in the change management pipeline.

The most popular process for the code review is that only a designated group of senior developers may review and approve the change. This approach has the limitation that it doesn't increase team familiarity with the codebase. If the group of final approvers is very small or consists of only the lead, it may also become a bottleneck. To overcome this limitation, the process may require another developer to review the code and agree or disagree with the change. Depending on the internal policy of the development team and the desired ratio of flexibility versus control, such peer review may be enough and will serve as the lead's approval of the change.

The pre-commit pipeline and code review process is supported in all major source code management systems, including GitHub, Bitbucket, and Gerrit. A similar process can be set up for non-Git-based source code repositories, like SVN.

The code review process is set up on all types of repositories, including business applications, system components, libraries, and tests.

7.4.3.2. Policy Inputs

Although the code review process should be set up by the release engineering team, the exact configuration of the policy should be supplied by the development lead. The inputs to the policy include:

1. List of people who can participate in the peer review.

2. Whether collective approval is allowed and, if yes, the number of positive reviews required to approve the change.

3. List of people who can approve the change after the code review.

4. Automated change validation rules.

7.4.4. Build and Packaging

The changes are first implemented in the source code. But before the source code can be deployed to production, it needs to be transformed into an executable code and packaged for deployment. The build and packaging process is executed on every change in the source code to ensure direct mapping between source code changes and deployment packages. Depending on the source code implementation technology, a build process may or may not be required. For example, server-side javascript, CSS files, and deployment scripts don't require compilation and building, although they still require packaging.

Build and packaging is a technical process, but it typically includes a number of validation and testing policies that we will discuss later. Those policies are defined and provided by the development lead. Together with code review, a successfully passed build and packaging stage is directly translated into the official sign off from the development lead.

The build and packaging process is different for libraries, business applications, system components, and tests. We will review these processes separately.

7.4.4.1. Deployable Units

First, let's review the concept of a deployable unit and deployment packages. Deployable units are read-only self-contained artifacts of an application or system component that can be directly deployed to an infrastructure-as-a-service by using provisioning and deployment scripts (Figure 7.11.). Deployable units are built once and can be deployed to any environment in the continuous delivery pipeline from test to production without change.

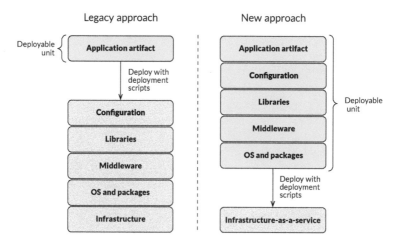

Figure 7.11. Traditional and current approaches to deployable units.

Traditionally, deployable units were only application artifacts and the final application was assembled only during the deployment phase. This was justified by the absence of infrastructure-as-a-service and the poor separation between infrastructure and applications. In the Java world, war and ear files were examples of deployable units that were linked at deployment time to the application server. This approach led to many configuration issues during deployment and familiar errors with environment instability or poorly reproducible errors when an application worked well in a test environment but misbehaved in production.

Self-contained deployable units decrease the number of configuration issues during deployment. To fulfill this requirement, deployable units need to include the application binaries, necessary configuration and properties, and all dependencies, including application libraries, middleware binaries, OS packages, etc. Because we require deployable units to be directly deployable to an IaaS, we will consider only two types: containers and VM images.

Deployable units are atomic and read only. It is not possible to build and deploy only a part of a deployable unit or to change a part of a currently running instance of a deployable unit.

7.4.4.2. Deployment Packages

A deployable unit is a self-contained but static artifact. In order to turn it into an instance of an application running on an environment, it needs to be deployed to an infrastructure-as-a-service with a deployment or provisioning script. Deployable units can be either VM images or containers, so their deployment on an IaaS is equal to provisioning resources; these two phases are therefore effectively joined into one.

Deployment units together with deployment scripts represent an executable deployable package. The package is produced from the same source code repository, so even if the build process produces different artifacts for deployable units and scripts, there is a strong bi-directional dependency between those artifacts. They are versioned together, so a certain version of a deployment script can deploy only a certain version of a deployable unit.

The deployment package represents the smallest granularity of service change after the build stage. Any change in deployment scripts, application code, application-specific configuration, or middleware leads to the creation of a new version of a deployment package. This new version then goes through a change management process before deployment to production.

A deployment package is executable and self-contained, so it needs very little to deploy an instance of a component. To execute, it only needs information about the environment, which includes endpoints and credentials to access the infrastructure API, cloud projects, and networks, service discovery endpoints, etc. Provided with this information, deployment scripts from the package use the infrastructure API to provision resources and run application instances (Figure 7.12.).

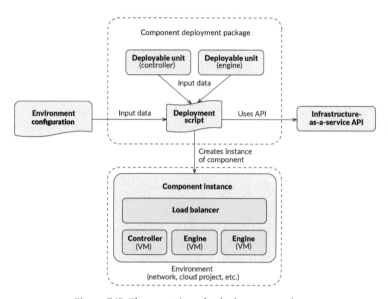

Figure 7.12. The execution of a deployment package.

The deployment package may be composite and may contain one or many deployable units. If a component or application is complex and consists of a number of different sub-components, the build process can produce multiple deployable units. For example, a complex component can have tightly coupled controller and engine instances that should always be released together. A deployment package may also serve as a packaging mechanism for all components in a service in cases when they should be versioned and released together as a single unit.

The key property of a deployment package is that it has a single handle to select and invoke. It may contain an immutable composite of any number of binary objects (such as VM images, configuration bundles, etc.) under the hood. Figure 7.12. labels this handle "deployment script" but it may or may not be a "script."

Let's review different build and packaging processes for different entities, including libraries, business (custom) applications, and system components.

7.4.4.3. Libraries

The build process for libraries is the simplest one because libraries don't produce deployable units and packages. Each commit to the source code repository with a library triggers a build process (Figure 7.13.).

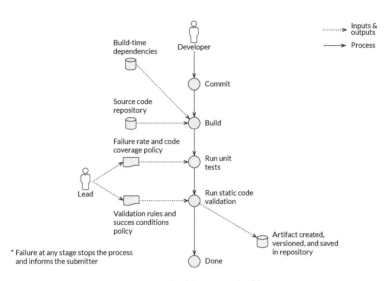

Figure 7.13. The build process for libraries.

Although libraries don't produce deployable units, they still produce artifacts that can later be referenced and used in business applications and system components. These artifacts need to be versioned and saved into a repository for consumption.

The build process for libraries may require other libraries' artifacts in the form of build-time dependencies during the build process. In this case, as in the case when applications and system components depend on libraries, only specific versions of dependencies should be used.

7.4.4.4. Business Applications

The build process for business applications is more complex, because they need to produce self-contained deployment packages with all dependencies and deployment scripts (Figure 7.14.).

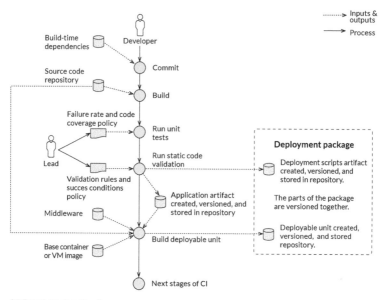

Figure 7.14. The build process for business applications.

The first part of the business application build process is similar to that of the library build process. The second part is different when the deployable unit needs to be created with all dependencies to middleware and a base image of a container or VM. Technically, middleware and the base container or VM images should be treated in the same way as application-level libraries. However, they are typically referenced and used at a later stage of the build process, when the actual deployable unit artifact is created.

7.4.4.5. System Components

The build process for system components (Figure 7.15.) is conceptually the same as the build process for business applications. Its final result is a deployment package, and it has some build-time dependencies. However, there are practical differences, because system components typically don't have any custom code apart from deployment and configuration scripts. The component binaries are pre-built and stored in an artifact repository. For example, it is not required to build MongoDB or Cassandra from sources every time. Their binaries are available for packaging.

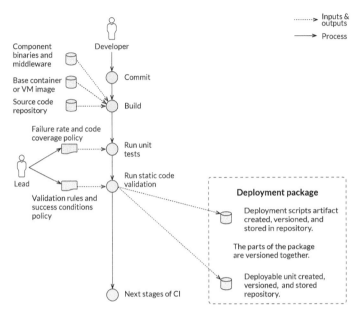

Figure 7.15. The build process for system components.

The system component policies for unit tests and code validation cover only deployment scripts. Often deployment scripts don't support unit testing and static code validation, and in this case, those policies will be empty.

7.4.4.6. Tests

Building tests produce a special kind of deployable unit. Tests are not deployed to production and are not subject to the change management process. However, functional and performance tests need to be executed against a service running in an environment. We can consider that tests are deployed and executed from a platform-as-a-service. For example, a combination of Jenkins with a Selenium cluster may serve as a PaaS for functional tests based on Selenium.

However, in the absence of a PaaS, tests can also be built into self-contained deployment packages that can be deployed directly on the IaaS. An example of such tests may be stability or performance tests. If tests are deployed into a PaaS, the build process for the tests will be similar to the build process for libraries. Otherwise, it will be similar to the business applications build process.

7.4.4.7. Versioning

Artifacts, deployable units, and deployment packages are created from specific changes in a source code repository and need to be versioned. All cases of build-time dependencies should reference specific versions of artifacts. For example, in the case of using the Maven tool to build Java code, referencing to "snapshot" versions should be prohibited. The use of specific versions is required for multiple reasons:

1. To trace changes in artifacts back to the specific commits in the source code repository.

2. To reliably debug and troubleshoot defects found during testing.

3. To have a history of past versions of artifacts for the audit trail.

It is important to note that artifact versioning is different from service API versioning. Service API versioning pursues a different goal of maintaining the backward compatibility of a service contract, even if the actual API contains backward-incompatible changes. For example, if a service is exposed over a REST protocol, its API contains the method /service_a/v1/sample_action, and the new version of the method is not backward compatible with the old one, maintaining the original method and exposing the new one at /service_a/v2/sample_action maintains the backward compatibility of the service API as a whole. Technically, the addition of a new version to the service API is similar to the addition of a new method, and additions are backward-compatible changes.

Unlike artifact versioning, API versioning rarely uses full semantic versioning and oftentimes works with only major and minor versions. Whereas artifact versions are unique and are incremented for every change, service API versions change rarely, typically when changes in the API are big and backward compatibility can't be supported.

7.4.4.8. Policy Inputs

When a change successfully passes the code review and build stages, it automatically receives a sign off from the development lead. To ensure legitimacy of the sign off, the development lead is responsible for providing a number of executable policies to the build process:

1. Unit tests if any.

2. Unit test analysis and success ratio. The most straightforward policy is to require all unit tests to be passed and have 100% success ratio. This is not always feasible though and may lead to frequent false negatives and later manual overrides of such false negatives by the development lead. To avoid this, the development lead may require a different policy with unit tests split into categories from most important to least important. For example, 100% success ratio for important tests and 90%+ success ratio for the least important tests may be enough to pass the build.

3. Minimum unit test coverage. We want to caution against using too high numbers for the test coverage, which may lead to the creation of unnecessary tests that are expensive to create and maintain but do not increase the overall quality.

4. Static code validation rules if any. The development lead has to choose whether to use static code validation, what tools to use, and what configuration of the tools to use. By default, static code validation tools have a strict configuration and produce a lot of errors and warnings. The development lead needs to choose a specific configuration of static code validation tools that is relevant to the application and then define a success ratio with the maximum number of errors and warnings allowed to pass the build.

During the build process, all outputs and policy execution results should be recorded in a database and tied to a specific change for the audit trail.

7.4.4.9. Example: eCommerce Platform

The eCommerce platform has quite a few of components, but many of them are similar from the build process perspective. We'll review several typical cases: the catalog business application, base Cassandra library, and checkout Cassandra component.

For this example, we will use a Docker and Kubernetes stack for packaging and deployment. However, other technology stacks, like VM images with Hashicorp tools and any mature cloud provider, can be used instead.

The catalog business application follows the standard business application build process, with a build-time dependency on a back-end library (Figure 7.16.).

The catalog application is implemented in Java, requires Jetty as an application container, and is deployed on a CentOS operating system. Its deployable unit is a Docker container, and the deployment script is implemented with Kubernetes and Helm manifests. The Docker container is built in two phases. First, an application artifact is built with Maven and the JAR file is stored in Nexus. Then, a deployable container is built by using Dockerfiles from the artifact, middleware dependencies, and a base container image. As discussed in the source code repository section, all build and deployment scripts are located in the same application code repository. When the Kubernetes and Helm deployment scripts are created, they are pre-configured with the same version of the Docker deployable unit. This allows the whole deployment package to be represented with the deployment script, because there is a strong dependency to a specific version of the deployable unit.

The base Cassandra library is maintained in a separate source code repository to facilitate reuse between the checkout and customer instances of Cassandra. It is a system component library, so it uses a slightly modified library build process (Figure 7.17.).

All of the source code for the base Cassandra deployment is in Dockerfiles, properties, Kubernetes manifests, and Helm charts, so we do not perform unit testing and static code validation. The resultant deployment scripts and Docker containers are not deployable units. They need to be extended with a specific configuration in order to become deployable units.

The customer and checkout Cassandra repositories contain a specific configuration of Cassandra cluster for those services. In the simplest cases, specific configurations will not build new containers and will use base ones. However, they will take base deployment scripts and override them with specific configurations. The build process for the checkout Cassandra is shown in Figure 7.18.).

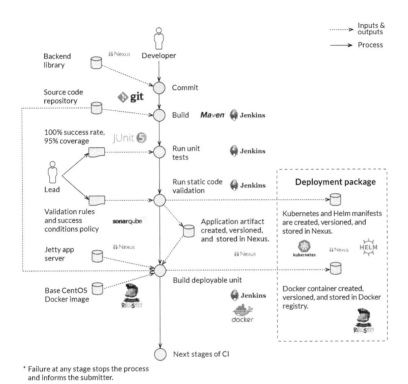

* Failure at any stage stops the process
and informs the submitter.

Figure 7.16. Build process for the catalog business application for our eCommerce platform example.

Because there is no application source code, the build pipeline for checkout Cassandra is shorter than for the catalog application. Although unit testing may be applied to Dockerfiles and Helm charts, we do not use it in this case, because these parts change rarely, they are simple, and successful deployment can be used as a sufficient test. As we decided to skip unit tests, there are no relevant policies, which simplifies the pipeline even further.

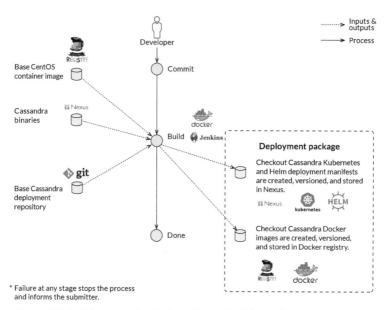

Figure 7.17. Build process for the base Cassandra library for our eCommerce platform example.

The checkout Cassandra pipeline leads to the creation of a deployment package that is similar in structure to the catalog application deployment package. When the base Cassandra repository is changed, it doesn't trigger a rebuild of the checkout and customer Cassandra packages. In order to get the latest changes from the base Cassandra repository, the checkout and customer Cassandra repositories need to reference the new version of the base Cassandra artifacts. That requires a commit into the respective repositories. This may slow down the process of rolling out enterprise-wide Cassandra changes, but it will help with higher visibility and control in how these changes are rolled out and will lead to higher stability of the system. It will also ensure that services consume only the changes that they need from the library components and don't get unneeded changes automatically without proper testing and verification. For example, if the base Cassandra repository migrates to a new version of Cassandra, it doesn't mean automatic rollout of the new version to all services. If the checkout service needs the new version, it will migrate. If the customer service doesn't need it and, in fact, that new version may break functionality, it can remain on the old version.

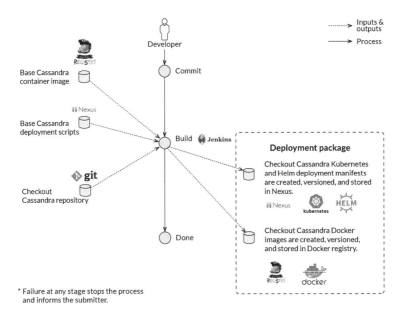

Figure 7.18. Build process for the checkout Cassandra component for our eCommerce platform example.

7.4.5. Deployment

After the deployment packages with the change are prepared, they need to be deployed to an environment. The deployment stage itself doesn't have any policies attached to it, but it is necessary to prepare the right environment to reliably test and certify the change before it goes to production.

As we described in the previous section, deployment packages are self-sufficient. In order to deploy component instances, they only need information about the environment:

1. Endpoints and access credentials to the infrastructure or platform API.

2. Information about the cloud projects and subnets on which to deploy the application.

3. Runtime scalability configuration information if the application is horizontally scalable.

4. Endpoint and access credentials to the service discovery and secret management service.

All application properties, including the size of VMs, settings of an application server, and garbage collection information, should be included in the deployment package. This will help to test applications in low-level environments in exactly the same configurations as they will be deployed to production. The exact split between application and environment properties may vary by application-level and organization-level change management policies and the level of tolerated risk. The general guideline is that the component should be deployed in the same configuration across environments to guarantee reproducible results. We will discuss below specific guidelines about what properties should be part of a deployment package, what should be part of an environment, and what can be configured by business users on the fly.

7.4.5.1. Deployment Policy

The requirements for deployment come from the testing policy. Specifically, the change should be tested in a non-production environment in exactly the same configuration as it would be applied in the production environment. The challenge with this requirement is that we need to guarantee that the configuration of the non-production environment is exactly the same as the configuration of the production environment will be after we have applied the change to it. In the most generic case, this means that we need to guarantee that:

1. If the current state of the production environment is A, we need to start with the state of the non-production environment being equal to A.

2. We need to apply the deployment package representing the change to the non-production environment in state A. The deployment package will transition the state of the non-production environment from A to B.

3. The state of the production environment shouldn't change between the time the change was certified in the non-production environment and the time the change was deployed to the production environment.

4. After the change is certified in the non-production environment, exactly the same change should be applied in exactly the same way to the production environment.

If all conditions are met, there is a guarantee that the final state of the production environment will be equal to B, which is the same as the result of the test in the non-production environment.

Unfortunately, these conditions are too strict and hard to achieve. The biggest problem is that the non-production environment can never be exactly the same as the production environment. It will have different networks, different data, different external dependencies, etc. So, the practical requirement is that the state of the non-production environment should be indistinguishable from the production environment from a service and change perspective. One of the easiest constraints to relax is the exact equality of data. Because services should be able to operate with any data within predefined boundaries, datasets in the production and non-production environments can be different. Another relaxation is to separate the configurations (properties) into application-level, environment-level, and business-level configurations and then allow the environment-level and business-level properties to differ between environments. However, the separation between different levels of properties should be done carefully to avoid changing application-related configurations between environments.

After relaxing certain requirements, we still need to have a non-production environment in state A', which is indistinguishable from production state A, before deploying the change. We also need to ensure that deployment of change X, which transitions the non-production environment from state A' to B', will transition the production environment from state A to state B and that the new state B will be the desired state. One typical challenge, however, is that the non-production environment may go through a number of state changes and, strictly speaking, change X is a composite change. For example, let's assume that there is a service of version v1 deployed to both production and non-production environments. We then deploy a service of version v2 to the non-production environment, but it fails testing. After fixing the defects, we deploy version v3 to the non-production environment and the testing is successful. We then deploy version v3 to the production environment on top of v1. However, there may be a problem here. From the non-production environment perspective, we deployed two separate changes, with the second change (v3) on top of the first one (v2). Deployment of a single v3 change to the production environment on top of v1 may lead to misconfiguration. This situation will lead to issues in cases when the deployment of changes is not associative and is not idempotent, which is typically the case when a deployment has side effects. Fortunately, if we follow the best practices of creating self-contained deployment packages and deployable units, as discussed above, the deployment of changes will be associative and idempotent in most cases. For example, when deployment is done by copying files or recreating VMs and containers, the changes are associative and idempotent. Database schema changes, by default, don't have these properties unless they are packaged and deployed correctly. For SQL databases, tools like Liquibase and Flyway solve the problem. If the deployment involves non-standard and configured external services, you need to pay attention to the right packaging and deployment of changes so that they satisfy the associativity and idempotency properties.

The requirement to retain the state of the production environment between testing of the change in the non-production environment and application of the change to the production environment is rather strict if the state of the production environment includes the state of all services and components running within that environment that are dependencies of the service under change. Typically, this requirement can be relaxed and other services can be changed if there is a guarantee that the changes to other services are backward compatible. In a good service-oriented architecture, this is a feasible requirement that allows the release of different services independently. In a transition from a monolithic architecture to a service-oriented or microservices architecture, this requirement tends to be hard to follow initially. It requires initial implementation of a global version. We'll discuss this topic in detail in a later chapter.

To summarize, the practical policies related to deployment are as follows:

1. Before deployment of a change to a non-production environment, the environment should be in a state that is indistinguishable from that of the production environment.

2. Exactly the same change that was deployed and tested in the non-production environment should be applied to the production environment later without modification.

3. The production environment shouldn't change in a distinguishable way after the change was deployed to the non-production environment and before it is deployed to production.

4. The deployment process that is used to deploy a change to the non-production environment should be the same for deploying that specific change to the production environment later. For example, if a blue-green upgrade (see the section later in this chapter) is planned to be used in production, the change should be deployed to the non-production environment in blue-green mode as well.

7.4.5.2. Application and Environment Properties

The separation between the application and environment properties serves the one major purpose of allowing us to reliably test a change in a non-production environment and have a reasonable guarantee that the change will work well after deployment to a production environment. This goal is often called the "build once, deploy anywhere" principle. In the previous section, we discussed that production and non-production environments should be indistinguishable from the application perspective. On the other hand, some properties need to stay configurable and allow a lighter change management process.

To achieve that goal, there should be a strong contract between the application and environment. In this case, by application, we mean a deployment package representing a service, business application, system component, or any other component that is under a change management process. By environment, we mean a representation of the rest of the world from the application perspective, where a deployment package is instantiated and an application instance exists.

The environment should provide a well-defined contract with a set of interfaces to an application. Figure 7.19. shows some examples of such interfaces. The interfaces should be the same across environments, but the implementation may be different within the boundaries of the contract. There are several main interfaces between an application and the environment:

1. Infrastructure. Different environments may use different cloud projects, networks, and credentials to provision resources for

applications, but the infrastructure API across the environments should stay the same.

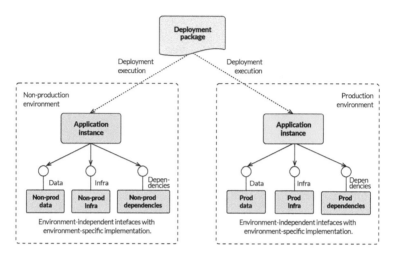

Figure 7.19. Interfaces between the environment and the application.

2. External dependencies. Whether dependencies are discovered dynamically via service discovery or configured during deployment, the actual instances of upstream services may be different across environments. The APIs and behavior of those service instances should remain the same across environments. (Here and elsewhere in the book, we use the term upstream to refer to services that are depended upon; in this case, that is, any services that the application depends on. Conversely, downstream services are those that are dependent on others.)

3. Data. Non-production and production data are different, but they should be within the boundaries allowed by the application. These boundaries are typically enforced by application tests.

4. Business configuration, including feature flags. This configuration is considered to be the same as user-generated data from an application perspective. We will discuss configuration changes in detail later in the book.

The first two interfaces are exposed via existing IaaS or services APIs and are completely independent from the application.

The latter two interfaces are more complex, because they are a part of the application and there is no special API for them. In order to ensure that the application can work well with various sets of data and feature flags, it should be tested appropriately. Specifically, if an application has two feature flags and it is known that those flags can be turned on and off in any combination in production by business users, the application tests should cover all four combinations of feature flags. The same applies to datasets. Various datasets are typically tested in the non-production environment to ensure that the application can work with any allowed data that can possibly appear in production due to the interactions of end users, business users, or downstream services with the application in question.

Some environment interfaces fall into a gray area and whether they belong to an application or the environment depends on the application, the level of testing that the team is ready to perform, and the level of risk that the release engineering and operations team is ready to take. For example, let's consider vertical and horizontal scalability configurations.

In the case of stateless applications that are embarrassingly parallel, the horizontal scalability configuration (the number of application instances) can be an environment interface. Such applications can typically scale in and out without any effect on their functionality. Of course, there are practical limits to which even embarrassingly parallel applications can scale. Those limits need to be tested or at least projected by the development team. They need to be explicitly declared in an auto-scaling policy or communicated to the operations team. For example, if an application is known to work well with the number of instances between 1 and 100, it shouldn't be possible to scale it out to 150 instances in production without additional testing or approvals.

Horizontal scalability works differently in the case of clustered applications, like Zookeeper, Cassandra, or MongoDB. Conceptually, such applications can scale horizontally within large boundaries, but practically, their specific configuration can reduce those boundaries significantly. If dynamic scalability for these applications is required in production, various cluster configurations and sizes should be explicitly pre-tested and declared as allowed ranges of configuration changes. The more simple and straightforward approach is to consider the scalability settings of a stateful clustered application to be part of the application itself. Then, a deployment package can be created with a fixed cluster size and configuration; when scaling is needed, it will be considered as a code change, which will produce a new deployment package that will be retested.

Vertical scalability is more challenging and constraining. Deployment of an application to a larger VM or container doesn't necessarily increase application performance. In fact, it can lead to crashes if the application itself or its middleware is not configured to utilize the new resources correctly. To avoid misconfigurations in production, dynamic vertical scalability should be avoided. The requirement for larger server capacity should lead to the development of a new change, and this change should go through the usual change management process. In those rare cases when dynamic vertical scalability is required in production without going through a typical change management process, various vertical scalability configurations need to be explicitly tested and different deployable units need to be pre-created as a part of the deployment package.

In general, the more configuration flexibility is required in production, the more testing needs to be done in pre-production. With a fast and efficient change management process, the overhead on testing may not be cost efficient and it may be easier and safer to consider some configuration changes as regular application code changes.

7.4.5.3. System Components

Before we get to the deployment methods, we need to review an important aspect of packaging and deploying business applications and system components. Earlier in the book, we defined services as a means to provide business functionality over a well-defined interface and as consisting of a number of business applications and system components. Typically, business applications contain business logic and expose it over a well-defined API, whereas system components store and serve data that belong to business applications.

System components, like MySQL, Cassandra, Redis, or Kafka, can also be thought of as services with well-defined APIs but with a very specialized interface. Changes to system components are expensive, mostly because they store user-generated data. This is particularly the case if these data are a system of record data that cannot be lost and reloaded from another source when a new change is deployed to production. Maintenance of a reasonable isolation between business applications and system components helps to limit the scope of certain changes to system components (Figure 7.20.). For example, changing database schema shouldn't always lead to the creation of a new system component deployment package with later redeployment of the system component instance.

Fortunately, the right degree of isolation is already provided by the system components.

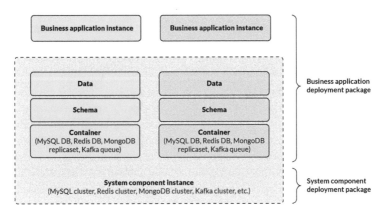

Figure 7.20. Isolation between business applications and system components.

The system component deployment package should contain only the middleware and configuration of the cluster that is specific to the business application. In turn, the application deployment package should contain the following, in addition to the actual application:

1. Deployment and configuration instructions for a data container. Most databases, caches, and message queues provide functionality to create several containers as a topmost degree of isolation. An example of a data container would be a MySQL database or MongoDB replica set.

2. Definition of a schema within the data container.

3. Pre-defined reference data stored in the system component.

The separation above means that when a system component is deployed, it doesn't have any explicit application-specific information, except for the non-functional configuration required to serve use cases of that specific application. Containers, schema, and reference data are created when a business application is deployed. If a change in schema is required, the system component shouldn't be changed. The change should be defined in a business application repository, which will lead to building of a new business application deployment package. The schema will be changed in a system component on deployment of the new version of the business application. Inside the business application package, the schema can be defined within the application code and deployable unit or within deployment scripts. Both approaches are acceptable. Embedding of schema updates into deployment scripts will lead to more control during deployment, but it may lead to a new version of a business application expecting a new schema and not finding it as a result of the asynchronous nature of the deployment process: for example, if a deployment script first instantiates a new application cluster and then executes a schema update but the application instances start and connect with the database before the schema update is executed. Addition of schema updates to the application startup may be beneficial if the technologies allow this to be done reliably and with a high degree of transparency. If schema updates are embedded into the application code, the technology should support idempotent schema changes and necessary synchronization mechanisms to ensure that multiple application instances starting at the same time and trying to update schema won't corrupt the schema during startup.

Even with the right separation between business applications and system components, changes to applications may require dependent changes in system components. For example, if a development team implements a change in the application that will increase the amount of data stored in the system component tenfold and change data access patterns, the configuration of the system component needs to be adjusted appropriately. In our experience though, such disruptive changes represent the minority of all business changes.

If a system component is not intended to be shared between services, the system component instance and container are tightly coupled. In this case, it may make sense to move the definition of a data container to the system component package and leave only the schema definition in the business application package.

When a microservices architecture is used, the general recommendation is to avoid shared system components. Dedicated component instances simplify the service-level release process because a new change affecting one service can be deployed without testing a global version of all services in combination. We recognize that, in some cases, dedicated components are not always practical or feasible. Shared components help with resource utilization and team efficiency. When done right, shared components may enable a partial platform-as-a-service.

Implementation of separation between the business application configuration and system component configuration, which was described above, helps with the implementation of shared system components. However, it is often not enough. The implementation of a shared system component requires additional design work and capacity planning to be done to ensure that the shared component can support multiple business applications and can onboard new ones dynamically. This additional effort typically makes implementation of true PaaS prohibitive and leaves shared components only for the most simple use cases. Such cases may include shared message queues or database clusters for several services within one business domain when the load from business applications is not significant.

7.4.5.4. Deployment Process

New instances of applications can be deployed from scratch or upgraded from the previous versions. The production environment is not transient, so all deployments to production are upgrades except for the initial one. To satisfy deployment policies, non-production environments need to have the same state as production before the deployment of new changes. This means that the majority of deployments to non-production environments are upgrades as well. Having said that, the creation of new application and service instances from scratch is still required for the initial setup of an environment. When dynamic environments are used for testing, the initial setup of environments may be frequent.

We will next review the recommended pattern and details of how the deployment process should be implemented. This process facilitates the separation of application deployment packages and environments, and it supports the implementation of deployment policies.

No matter whether an application instance is deployed from scratch or upgraded, the same base process is executed. The deployment process has several inputs:

1. Deployment package with deployment scripts and deployable units.

2. Environment configuration, typically in the form of properties.

3. Infrastructure-as-a-service API, be it a cloud or container management API.

The deployment scripts use the IaaS API to provision deployable units in accordance with environment properties. Because the deployable units are VM images or containers, provisioning of resources is the same as deployment of an application. Note that in the process we describe below, deployment scripts do not provision VMs or containers directly. Instead, we recommend decoupling provisioning and configuration of application templates, in the form of lifecycle management controllers and image templates, and actual provisioning of VM or container instances. This separation improves the management of deployed applications after they have been deployed by enabling automatic execution of auto-scaling and self-healing policies in runtime long after the initial deployment is done.

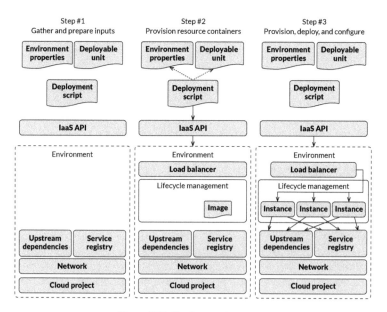

Figure 7.21. Deployment process.

In the first step of the process (Figure 7.21.), all necessary inputs are gathered. The deployment script finds deployable units, gets access to the IaaS API, and finds the proper environment for deployment of the component.

In the second step, the deployment script uses the IaaS API to provision base resources and conFigure resource containers. The deployment script doesn't provision VM instances or Docker containers directly. Instead, it uses a VM image or Docker container from the deployable unit to create a resource lifecycle management controller in the IaaS. All modern IaaS providers provide such capabilities. For example, Kubernetes provides deployments and replication controllers. Google cloud platform provides managed instance groups. AWS EC2 provides auto-scaling groups. These tools help to manage the lifecycle of application instances, including deployment and provisioning, self-healing, auto-scaling, and registration with load balancers. The deployment script configures the lifecycle management controller with templates of the VM images or Docker containers, adds environment-specific configuration to the templates, and instructs the controller with the minimum number of instances to create initially. The environment-specific information typically consists of the environment properties passed to the deployment script and includes the endpoint and credentials to access the service registry, endpoint and credentials to access the IaaS API, etc. In addition to the lifecycle management controller, the deployment script creates load balancers, firewall rules, storage objects, and any other IaaS-native objects that are a part of the deployed service.

After the completion of the second step, there are no VM instances or containers running with the application. Only the template of the service is provisioned and configured.

In the third step, the infrastructure provisions the required number of VM instances or containers. On the start of the operating system, the middleware and the application inside are launched. During the launch, either the application itself or helper scripts find the service registry [33] and perform a number of registration and discovery steps to conFigure the application with its dependencies:

1. The service endpoint of the newly deployed application is registered in the service registry for later consumption.

2. The endpoints of upstream dependencies are discovered and passed as configuration to the application.

Depending on the implementation of the service registry and whether the service registry is pre-integrated into the IaaS, the steps above can be performed by the IaaS itself, the application code, or helper scripts. For example, if Netflix Eureka is used, the application code will perform registry and discovery functions. If Hashicorp Consul is used, its configuration will do the work. If Kubernetes is used, registry and discovery will be performed by the platform.

After the startup, the application also needs to get the business configuration, like feature flags. Sometimes the business configuration can be stored in the service registry, but it is most often done by a separate system. We will discuss business configuration and the associated changes later in the book.

In most cases, the service registry stores only non-sensitive data. However, most system components and some business services may require authentication information. Some components may also require secure sockets layer (SSL) certificates. This information is typically stored and extracted from a secret management service. The protocol for extracting this information is similar to the one described above for the service registry. Some platforms, like Kubernetes, have secret management pre-integrated. Otherwise, tools like Hashicorp Vault can be used.

Once the third step is complete, the application instances are started and configured, and the health and readiness checks are passed, the service is ready to serve traffic. The VM image or Docker container templates remain in the lifecycle management controller though, so if one of the instances needs to be shut down and reprovisioned or the set of instances needs to be scaled in or out, the controller will use the template to provision new application instances as it did during the initial deployment.

7.4.5.5. Rolling Upgrades

After the initial deployment of the service instance is performed, subsequent changes are implemented as upgrades. There are two popular upgrade procedures that are used most often: rolling upgrades [34] and blue-green upgrades [35]. Rolling upgrades are in-place upgrades, when the old instances are replaced one by one with the new ones. It starts with the previous version of a service deployed. Let's assume that a service consists of a business

application behind a load balancer with a database system component (Figure 7.22.).

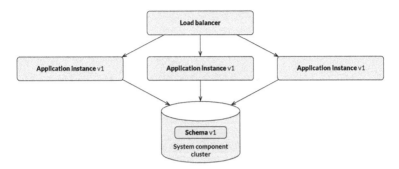

Figure 7.22. Original version of a service before a rolling upgrade.

The next step is to deploy a new version of an application instance
alongside the existing version (Figure 7.23.). When the new version
of an application instance is deployed, it connects to the same load
balancer and updates the schema and data in the database system
component. To avoid issues with the previous version of the application, the schema and data changes have to be backward compatible so that the application of version v1 can still work with the
schema of version v2.

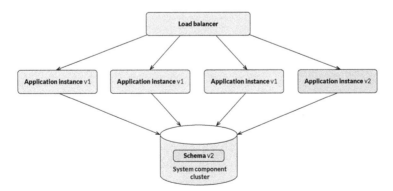

*Figure 7.23. Version v2 of an application instance is deployed alongside
version v1.*

When the application instance of the new version is ready to serve traffic, one of the old instances is destroyed and another instance of the new version of the application is deployed (Figure 7.24.).

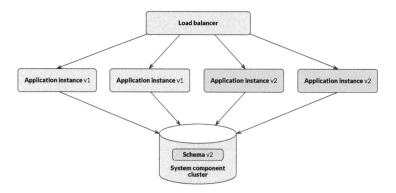

Figure 7.24. A version v1 instance is destroyed and replaced with another version v2 instance.

Ultimately, new versions of application instances replace all of the old ones (Figure 7.25).

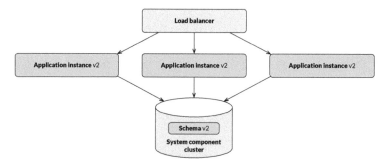

Figure 7.25. Version v2 instances replace all of the version v1 instances.

Most infrastructure-as-a-service providers, including container management systems, support rolling upgrades out-of-the-box. They are typically implemented by replacing the old version of the deployable unit template in the lifecycle management controller with the new one. The controller then will gradually replace old instances with new ones. The exact configuration of the upgrade is adjustable:

1. The step of the upgrade can be specified; that is, the instances can be replaced not one-by-one but in sets of two, three, or any other arbitrary number. This helps to finish upgrades sooner.

2. The maximum and minimum number of old and new instances can be set, so that the cluster size doesn't become too big or too small during the upgrade.

Rolling upgrades have a number of important caveats. Application instances are upgraded in place under the same load balancer, so different versions of an application are available to the downstream systems at the same time. Because the use of sticky sessions on a load balancer is not recommended, it also means that one logical session can be served by different versions of an application. Therefore, the new versions are required to be backward compatible with the previous ones. Rolling upgrades are typically used for minor application changes or when blue-green upgrades don't work well.

In our experience, one of the common use cases for rolling upgrades is upgrading stateful applications with a system of record data, like databases. Most stateful applications are clustered, requiring a very specific number of instances with well-known addresses running at the same time. Zookeeper and MongoDB are examples of such applications. Zookeeper clusters have to have an odd number of nodes in order to ensure quorum, so new instances cannot be arbitrarily added to the cluster. In these cases, rolling upgrades would replace instances one-by-one. Old version instances are terminated first, then new version instances are created with the same well-known addresses and added to the cluster. For an example, you can look at how Kubernetes deals with rolling upgrades to stateful sets [36].

7.4.5.6. Blue-Green Upgrades

Rolling and in-place upgrades are supported by the majority of modern deployment tools and infrastructure-as-a-service, which makes them the simplest forms of upgrade. A very small amount of custom orchestration is required for executing these upgrades, making them very fast and simple. However, sometimes additional controls and orchestration are needed to:

1. Ensure that new versions work well and that they can success-fully pass complex functional and nonfunctional testing, in ad-dition to satisfying basic health checks.

2. Ensure isolation and separation of requests between old and new versions.

3. Enable canary releases [37] and A/B testing.

Blue-green upgrades satisfy all of the requirements above. In the case of a blue-green upgrade, a new instance of an application or component is created alongside the old one. We'll use the same ser-vice as we used for rolling upgrades as an example. The service con-sists of a business application under a load balancer and a database system component cluster (Figure 7.26.).

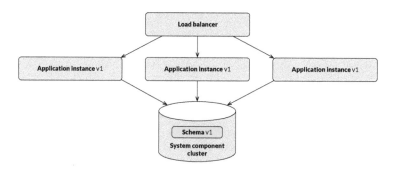

Figure 7.26. Original version of a service before a blue-green upgrade.

If there is a new version of a business application and the sys-tem component remains unchanged, the blue-green upgrade can be shallow or deep. A shallow upgrade will replace only application instances and will leave all other components untouched. The first step of the upgrade is to create a new instance of the business ap-plication (Figure 7.27.).

The new "green" version is deployed side-by-side with the old "blue" one. It is isolated from existing traffic but uses the same da-tabase system component. When the new instance of the business application is deployed, it upgrades the schema in the system component. This means that in the case of shallow blue-green upgrades, as in the case of rolling upgrades, schema changes should be backward compatible.

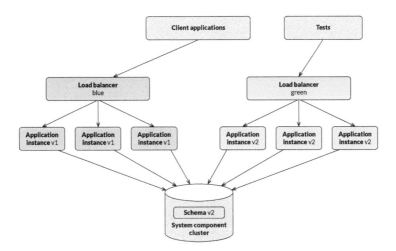

Figure 7.27. Version v2 of an application instance is created alongside version v1.

After deployment is complete but before the "green" version starts serving traffic from client applications, it is tested. The same system component is used for both new and old versions, so tests should be non-intrusive and shouldn't modify production data. After tests are completed, the traffic can be routed to the new instance. If there is no additional routing layer between the client applications and the service, the traffic can only be switched between 100% and 0%; otherwise, the shift can be made gradually. After the traffic is switched to the new version and old connections are drained, the old version can be destroyed (Figure 7.28.).

In this case, the switch is performed via either dynamically reconfiguring client applications with a service discovery mechanism or by switching DNS records to the new green load balancer. In both cases, the switching process is eventually consistent and draining the traffic from the old version can take some time.

A better implementation approach can be applied when the load balancing layer supports routing logic. Many application-level L7 load balancers, like HAProxy, F5, or AWS application load balancer, support such functionality. Most client-side load balancers, like Netflix Ribbon or linkerd, support routing as well. If routing is supported, a gradual switch of traffic and canary releases can be implemented (Figure 7.29.).

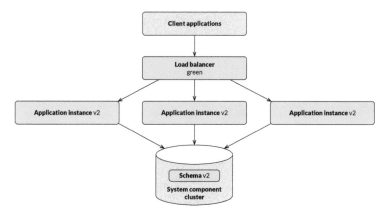

Figure 7.28. When version v2 is operational, version v1 is removed.

As we mentioned above, in shallow blue-green upgrades, only the application with changes is upgraded, which leaves both versions working with the same system component and same data over a period of time. In our experience, one of the common cases for shallow blue-green upgrades is when the system component of a service is a database that stores a system of record data. In this case, rolling upgrades or shallow blue-green upgrades of an application are the only options. However, when system components contain replicas of data and can upload data from the system of record at any time, it is safer to go with deep blue-green upgrades. In the case of deep blue-green upgrades, all components in the service are duplicated in the new version (Figure 7.30.).

In this case, after the new service instance is deployed, data need to be uploaded to the system component cluster.

As with rolling upgrades, blue-green upgrades don't always require the creation of a fully scaled new instance of the service right away. Initial deployment may create a small instance and scale it up only when requests from client applications start flowing into the new instance.

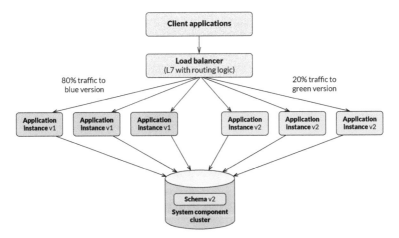

Figure 7.29. If routing is supported the switch from version v1 to version v2 can be performed gradually.

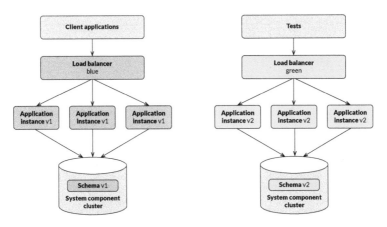

Figure 7.30. In deep blue-green upgrades, all components of the system are duplicated for version v2 alongside version v1.

Blue-green upgrades are supported by a number of deployment tools, as well as container management systems. Typically, blue-green upgrades are performed by deploying a new instance of the lifecycle management controller with a new version of the deployable unit template. When the traffic is shifted to the new version, the old controller is destroyed, together with old instances and templates.

However, blue-green upgrades often require custom orchestration. Decisions such as when to run tests and what tests to run, when to switch traffic to the new version, and how much traffic to switch depend on the particular service and change management policies. These decisions require custom orchestration and sometimes manual steps.

7.4.5.7. Upgrades Deep in a Service Mesh

The upgrades discussed above are base mechanisms for upgrading individual components and services. The real systems include multiple services communicating with each other, as in the example of the eCommerce platform we discussed above. There are special cases for how blue-green and rolling upgrades may work in cases of a service mesh [38]. Let's consider a simple system with three services working together with a top-level UI application (Figure 7.31).

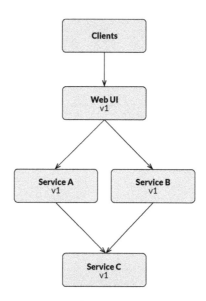

Figure 7.31. A simple example of a service mesh.

Rolling upgrades are the simplest option for upgrading any service in this hierarchy, because all endpoints and connectivity between the services will remain the same. Unfortunately, rolling upgrades are not always feasible if more control over the upgrade procedure is needed. This may happen when additional testing is needed before routing traffic to the new version or when canary releases are required.

Blue-green upgrades within service mesh hierarchy are more difficult to execute. The simplest blue-green upgrade case is performed on the top level services that are directly exposed to the clients. For example, if the web UI needs to be upgraded, the intermediate deployment topology will be as shown in Figure 7.32.

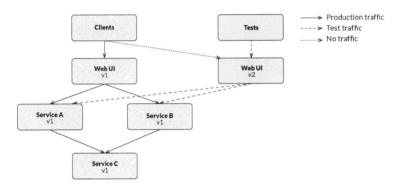

Figure 7.32. Intermediate deployment topology during a blue-green upgrade of the top-level web UI service.

The process is simple and resembles a shallow blue-green upgrade for a singular service. When testing of the new web UI version is complete, traffic can be switched to it. Because top-level services have well-known DNS names and are exposed to the outside world via rich application-level L7 global load balancers with routing logic, it is easy to implement traffic switching to the green version with various policies, including canary releases, different percentage allocations, and even A/B testing.

Blue-green upgrades of services deep in the service mesh are more challenging.

When a new version of service B is created, it needs to be tested. If service-level tests are enough, then the traffic can be switched after the testing is complete (Figure 7.33.). If the service B load balancer supports routing, canary releases and gradual traffic shifting can be performed as well. However, there is a problem if integration tests are required to certify the change or if canary releases can only be implemented on the top level. Integration tests and clients will continue working with the old instance of the web UI, but it somehow needs to be ensured that the tests and canary traffic are routed to the green instance of service B. One way to solve this problem is to implement routing fabric. We will discuss the routing fabric in the Further Improvements chapter later in the book.

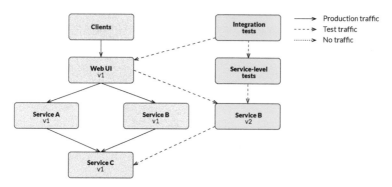

Figure 7.33. Intermediate deployment topology during a blue–green upgrade of service B embedded within the service mesh.

Another option to solve these issues is to perform a blue-green upgrade for all downstream services from the upgraded service. In this case, the intermediate deployment topology will as shown in Figure 7.34.

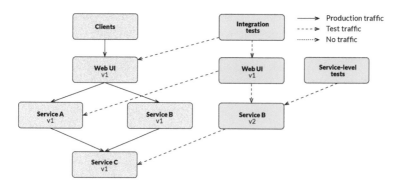

Figure 7.34. Intermediate deployment topology during a blue–green upgrade of service B and all downstream services in the service mesh.

In the case of upgrading service B, a green instance of the web UI service will need to be deployed. As the web UI doesn't need to be upgraded, the green instance will have the same version v1 as the blue instance. The green web UI instance is connected with the green instance of the service B component. Both service-level and integration tests can run on the new green service subgraph. After the tests have been successfully completed, client traffic can be gradually switched to the green instance of the web UI service, a process similar to the previous case when only the web UI service was upgraded.

Although this design solves some issues, it leads to a number of implementation complications. It is assumed that, during deployment, applications and services are using a service registry to discover dependency endpoints. In the case of blue-green upgrades within the service mesh, the service registry needs to support tags or labels to distinguish between different "colors" of the same service. For example, when the green instance of the web UI is deployed in Figure 7.34., it needs to connect with the green instance of service B. However, the green instance of service B is deployed, it needs to connect with the blue version of service C. At the same time, when the green version of service B is deployed, it needs to register its endpoint with a green label to distinguish itself from the existing blue instance of the service. The information about service colors and dependency colors needs to be supplied during deployment.

Another complication is that, after the blue instance is destroyed, the next upgrade needs to deploy the new version of the service with the blue label. To avoid confusion between colors, they need to be either swapped to the original state after the upgrade is done or the information about the currently used colors needs to be stored in an environment. Most service registry and discovery tools support setting labels on services, so it is possible to implement the right upgrade procedure with reasonable customization.

7.4.5.8. Readiness and Health Checks

Both rolling upgrades and blue-green deployment require the application to provide information about whether it is healthy and ready to serve traffic. During a rolling upgrade, this information is used to decide whether new instances are ready and a number of old instances can be shut down. During blue-green upgrades, this information can trigger functional test execution or switching of traffic to the green version of the service. The checks can be considered as lightweight tests ensuring that the applications are working correctly without running fully fledged functional tests or deducing this information from logging and monitoring tools.

Health and readiness checks are used for other features provided by infrastructure-as-a-service or container management platforms. Those features include self-healing and automatic registration and deregistration of application instances in load balancers. Two types of checks are needed, because they support different use cases:

1. Health checks show whether the instance of an application is working as intended and eventually will be able to serve user requests. If it is not healthy for a period of time, it should be destroyed and recreated.

2. Readiness checks show whether the instance of an application is ready to serve traffic. At some point in time, the instance may be healthy but not ready to serve traffic, because the caches are not warmed up or some preparation business logic is still running.

All components are generally required to implement both health and readiness checks. In practice, system components like databases may not implement readiness checks because they are not working behind load balancers.

7.4.5.9. Example: eCommerce Platform

Let's start by reviewing the deployment architecture of the eCommerce platform. For this example, we will assume that all components are deployed on Kubernetes in Docker containers (Figure 7.35.).

The eCommerce platform is exposed to customers via Akamai, which serves as a global traffic manager or global load balancer. The Akamai load balancer has rich routing capabilities and allows canary releases and gradual traffic switching between different versions of the web UI.

All business applications are running behind load balancers. When business applications are deployed, they register load balancer endpoints in the service registry. If one business application works with another service, the discovery is performed and the client application is configured with the load balancer endpoint of the

server application. Business applications are stateless and horizontally scalable, so they are deployed in replication controllers with
enabled auto-scaling.

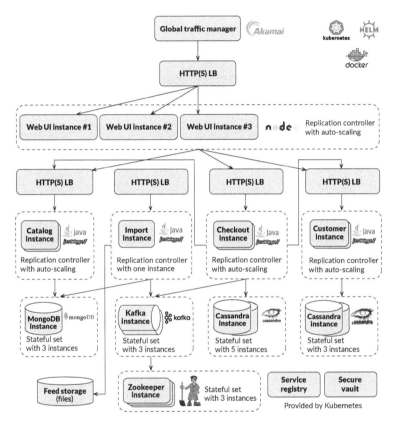

Figure 7.35. Deployment architecture for our eCommerce platform example.

System components are stateful applications, so they are deployed in stateful sets, which help with clustering and guarantee
the number of instances in the cluster. System components are not
exposed via load balancers. Instead, the business applications use
client-side libraries to work with the system components. The client-side libraries are configured with the endpoints of all instances
of system components.

All system components are dedicated to their respective service,
except for messaging component implemented in Kafka. Kafka is

used by two business applications: checkout and catalog import. In this case, this setup makes sense because the load on the Kafka cluster by both components is minimal and the level of isolation provided by Kafka between topics and queues is sufficient. Implementation of a shared Kafka cluster saves on infrastructure and maintenance costs. The separation of the schema and core component middleware is implemented according to the guidelines described earlier in the Deployment chapter. In the case of Kafka, the respective business applications create their own topics. Definitions of these topics are located in the respective business applications' deployment packages, whereas the Kafka system component remains generic.

From the upgrade perspective, all major changes in business applications are performed with blue-green upgrades. All changes in system components, except MongoDB for the catalog, are performed with rolling upgrades. MongoDB changes are performed with blue-green upgrades, because the catalog database doesn't contain a system of record data and can be populated with data by using the catalog import application. Minor backward-compatible changes in business application components and MongoDB are performed with rolling upgrades. Typical examples of such changes are bug fixes and minor new features that don't significantly affect business functionality.

For the purposes of this example, we will assume that the load balancers in the platform don't provide routing capabilities. The only load balancer with routing capabilities is Akamai. We also assume that top-level integration testing is required for any feature rollout, so blue-green upgrades of services in the middle of the graph require deployments of new versions of all of their downstream dependencies. When a new instance of a subgraph of components is deployed, new versions are tested. After successful testing, production traffic is switched to the new instance at the Akamai level. The implementation is similar to what was described above in the section about upgrades in service meshes.

Let's review the worst-case scenario of the MongoDB blue-green upgrade procedure (Figure 7.36).

In this case, new "green" instances of catalog, import, checkout, and web UI application components are created. All these instances are created because they represent a transitive set of downstream dependencies of the catalog MongoDB. New instances still have old versions of deployment packages, because they have not been not changed.

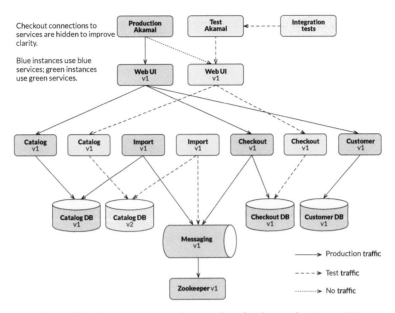

Figure 7.36. Blue-green upgrade procedure for the catalog MongoDB component of our eCommerce platform example.

In order to run end-to-end verification, a test Akamai endpoint is created and connected to the new instance of the web UI. This upgrade topology helps satisfy the requirement of running end-to-end testing before switching traffic to the new instance. In the absence of routing functionality in load balancers, it also enables canary releases and a gradual shift of traffic from the previous to the new versions.

In practice, this approach is not required in the majority of cases for several reasons:

1. The changes are done on service level, so service-level testing should be sufficient. If a system component in a service is upgraded, the only downstream application that is affected is a business application within the boundaries of this service. In other words, in the example above, only the new instances of the catalog and import applications will need to be deployed. Both the catalog and import applications need to be deployed in this case because they both belong to the catalog service. All downstream services and applications, like the web UI and checkout, are protected by the well-defined contract, so they

will stay as is and will switch to the new version of the catalog service.

2. Canary releases and gradual traffic shifting are not required, and traffic can be switched to the new version of the service immediately without affecting customers.

3. Many modern load balancers implement routing capabilities.

4. Small changes can be performed with rolling upgrades.

We decided to show this approach as an example, because it can be used to implement changes in an extremely safe and controlled way. Even the simplest technology stack would support this topology.

7.4.6. Service Testing

By the time a change is ready to be tested, it has been defined as a business requirement, has been developed and committed into the source code management, has been built and packaged into a deployment package, and is ready to be deployed into a non-production environment. At this time, the change is represented as a new version of a deployment package. Although the scope of each deployment package is a single component, the original change is defined as a business requirement on the service level. Therefore, a single business change may include one or many new versions of deployment packages that correspond to the same service.

The key goal of service testing is to receive a sign off from the QA team that the change is ready to be deployed to production. Depending on the exact organizational structure, there may be many sign offs from different sub-teams for different requirements: function, performance, stability, security, compliance, and others. Service testing should include all of these different types of tests, and all sign offs should be collected by the end of it.

At this point, we assume that the scope of a change doesn't span multiple services. Different changes in different services are implemented and can be released in parallel. Cases when changes in different services need to be tested and released together are discussed later in the book. Our assumption requires that changes are backward compatible from the service contract perspective. Specifically, new version deployment packages containing a change should be able to be deployed to production without breaking the downstream dependencies of existing versions . Backward compatibility is an important assumption that should be taken into account by the development team when implementing a change and should be verified during the testing phase.

The backward-compatibility requirement applies to the state when the new version of the service is already deployed and active. The transition from the state when the old version of the service is active to when the new version is active is not an atomic or transactional process. As we described in the previous chapter, this process is typically implemented with blue-green or rolling upgrades. Both methods have a transition period when both versions are active. The testing process should ensure that, during such a transition period, the old version of the service and the environment as a whole function correctly.

In the previous chapter, we discussed the deployment policies and requirements for environments at length. The goal of deployment is to ensure that the final state of the non-production environment after applying the change is the same as the final state of the production environment would be after that same change was applied. As the transition of the environment from the old version to the new version is important, the deployment methods for applying changes to non-production and production environments should also be indistinguishable. This specifically means that, if a change is deployed to a non-production environment with a rolling upgrade, it will need to be deployed to production with a rolling upgrade as well.

Testing is embedded into the continuous delivery pipeline, so both test execution and test results analysis should be automated. In this case, tests, test data, and test results analysis rules are the actual executable policies provided by the QA team. Execution of these policies provides a sign off that can be used to reject or approve the change.

7.4.6.1. *Testing Levels*

An original change is defined at the service level but implemented at the component level via deployment packages. Depending on the design of services and risk profile of a change, the QA team may also require integration testing of a new version of a service with other downstream and upstream services before they are able to sign off on a change. The choice of the right level at which to implement tests should find a compromise between the efficiency of the change management pipeline and the cost of creating and maintaining the tests.

Tests can be implemented on three different levels:

1. Components: A component can be tested in isolation from its dependencies, irrespective of whether it is a business application or a system component. For example, in Figure 7.37., a cache component of service B can be tested in isolation. Another more logical example from Figure 7.37. would be a messaging component, which doesn't belong to any particular service. The challenge with this approach is that individual components don't typically provide well-defined contracts for other components or such contracts may be too complex and difficult to test.

2. Services: Although a service is a logical entity, it has one or more components that actually provide the service API. Typically, such components are business applications. In service-level testing, all components of a service are tested as a whole, in isolation from other services. For example, in Figure 7.37., tests can be implemented to test service C as a whole by using its business application as an interface.

3. Groups of services: Several services may provide a certain functionality together. Sometimes, a top-level service exists that integrates all upstream services and provides an API or UI functionality for a group of services. In this case, a group of services can tested as a whole via this top-level service. These tests are typically considered as integration tests. In Figure 7.37., service A can be used as an aggregator and tests for all three services can be implemented at the level of service A.

We want to highlight the fact that unit tests are not considered to be one of the levels of testing. Unit testing focuses on verifying individual modules, functions, or classes in a source code. The granularity of unit testing is less than a single component and doesn't correspond to the granularity of change. Hence, unit testing is typically implemented by the development team and can't be used to validate a change on a service level.

There are several considerations that should be taken into account when choosing the right level of testing:

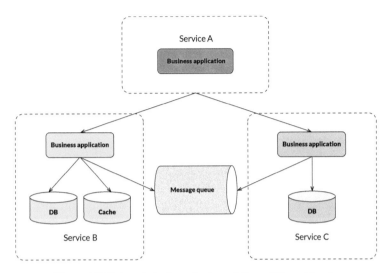

Figure 7.37. Setup that allows testing at three different levels.

- The amount and complexity of custom business logic. If a component or service doesn't contain significant business logic, the cost of testing it in isolation may be higher than the realised benefits. This typically applies to system components that have limited custom logic in the form of configuration. On the other hand, if a service group contains multiple services, each with complex business logic, the implementation of tests at the service group level may lead to an enormous number of test cases that are costly to create and maintain, difficult to debug, and slow to execute.

- The number of upstream dependencies. Because real instances of dependencies cannot be used for testing in isolation, more integration means more effort spent on implementation and the maintenance of stubs and mocks [39].

- The number of downstream dependencies. If a service or component is used by multiple downstream dependencies, it should be tested in isolation to avoid the duplicate effort of testing it in a group with each downstream dependency.

In our experience, when a system is designed according to service-oriented or microservices architecture principles, these considerations typically lead to the implementation of the majority of tests on a service level. This happens because services are initially designed to be reusable and have just the right amount of business logic.

If a service is tested in isolation, it typically leads to the creation of some kinds of test doubles [40]. However, a portion of service-level tests can still be done with real integrations. In this case, the cost of test creation and maintenance can be optimized further: service-level tests of downstream services can be used as integration tests for upstream services. For example, for the setup in Figure 7.37., service-level tests of service A can be used to test changes inservice B, in addition to service B tests. The same approach can be used for testing independent system-level components, like the message queue in Figure 7.37. Tests of service B and service C can be used to verify a change in the message queue, in addition to component-level tests. We understand that this practice may not be common, it would probably not be the default recommendation, and it needs to be approached carefully. However, we would still like to explore it in detail later in this chapter because it may help in lowering the barrier for the transformation to microservices architecture and continuous delivery.

The considerations for choosing the right level of non-functional testing are the same as for functional testing. In most cases, it makes sense to implement non-functional tests on the service level. However, non-functional tests are typically more expensive to create, maintain, and execute, so instead of the proper level of testing being estimated by the amount of business logic, it should be estimated by the complexity of the architecture and non-functional requirements. In some cases, it may be more beneficial to create non-functional tests on the service-group level.

7.4.6.2. Test Environment: Data and Services

Before the actual testing process can start, the test environment should be properly configured. In order for tests to be effective, the non-production test environment should be indistinguishable from production environment. As we discussed in detail in the chapter dedicated to deployment, there are four primary interfaces between an environment and a service: infrastructure, external dependencies, data, and business configuration including feature flags. The infrastructure and feature flags are covered in other chapters. In this chapter, we'll focus on the two pieces of environment configuration that are often the most difficult to implement correctly: data and external service dependencies. In our experience, the implementation of proper test data management, stubs, and mocks is the biggest challenge of test automation.

The implementation of data and external service dependencies in non-production environments has two major decision points (Figure 7.38.):

1. Whether to use production data or test data for testing a service.

2. Whether to use real instances of external dependencies or use stubs and mocks.

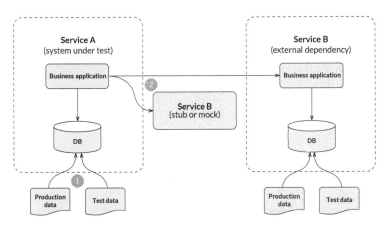

Figure 7.38. The two major decision points of data and external service dependencies in non-production test environments.

The easiest answer for both questions may be to deploy new version of a service directly to the production environment and connect with the production external dependencies without routing actual traffic through until all tests have been executed and sign offs received. This approach may radically simplify and improve the change management process, but it is not easy to implement correctly at the first attempt, and we wouldn't recommend it for teams that are just starting their journey toward continuous delivery. We will discuss this approach later in the book in the Further Improvements chapter.

A more typical case is when a new service version is deployed to a non-production environment. In this case, if a real external dependency instance is used, that instance also needs to be configured with test data, so the first decision needs to be made for the second service. From the other point of view, if a decision is made to use mocks or stubs in place of real dependencies, testing with real upstream services instances is still required to validate that integrations are working correctly.

There are several considerations when deciding whether to go with real integrations or service virtualization in the form of test doubles, such as stubs and mocks:

1. The cost of infrastructure and complexity of the environment. The most straightforward approach is to create test environments for each service, with dedicated instances of all dependencies down the stack. However, for complex service-oriented systems, such environments may turn out to be extremely large, which would increase the cost of infrastructure, reduce the stability of the environments, and lower the speed of creation of new environments.

2. The maturity of test data management. An alternative approach is to have shared instances of external dependencies. In this case, shared instances need to support different versions of downstream services, which, in turn, may have different requirements for test data in the dependency. Management of such test data without a mature test data management approach is difficult. The positive side is that mature test data management needs to be implemented in any case.

3. The cost of creation and maintenance of stubs and mocks. Implementation of service virtualization requires effort. Even with tools that allow the creation of stubs by recording requests and responses of real dependencies, the maintenance of proper versions of recorded data and management of dependencies between these versions and new versions is error prone and requires effort.

4. The complexity of the contract, complexity of dependencies' business logic, and side effects in external dependencies. The best case for stubbing an external dependency is when that dependency has read-only data and has a simple contract but

extremely complex internal business logic. Otherwise, the creation of mocks supporting modifications of data as a part of the contract may require duplication of business logic of the actual dependency.

5. The performance of external dependencies. If the performance of an external dependency is extremely slow, it may make sense to implement stubs for this dependency. However, slow responses will still need to be modeled in stubs for some test cases to simulate the actual behavior of the real dependency.

6. The quality and stability of external dependencies. If live external dependencies have poor quality and stability, the implementation of proper mock services may help finding root cases more easily.

Although there is no single answer for all cases of external dependencies management, it makes sense to use real instances of services whenever possible to keep non-production environments closer to production and avoid duplication of testing with stubs and with real dependencies. The use of this approach will help to close the gap between the test and production environments, minimize the number of test environments, and simplify and shorten the continuous delivery pipeline.

If the creation of test doubles is justified, there are several approaches to achieve that goal. A detailed discussion about the actual implementation of stubs and mocks is beyond the scope of this book. There are numerous articles on the subject and several open source and proprietary tools that help with the implementation of service virtualization.

The decision about production versus test data is oftentimes more straightforward. If the production data contain PII, PCI, or other sensitive information, the production data cannot be used for testing. In any case, there is typically no direct access to production data sources within a test environment, so the use of production data is an optimization technique to create robust test data cheaply.

Although the use of production data as test data can be efficient from the implementation perspective, it has several downsides. First, the size of the production data may be unnecessarily large. Copying of such data from the production environment, obfuscation as needed, and insertion into the test environment may take significant time and increase the infrastructure costs of the test environment. Second, a snapshot of the production data may not contain all combinations of data that are required to test all corner cases of the service. On the other hand, the latter argument applies to test data as well, and the use of production data may help in finding issues that were missed with test data.

In any case, tests need to support working with both test and production data, because the final smoke testing of a service is oftentimes performed when it is deployed to the production environment. Our recommended approach is to implement test data generation for non-production environments but implement an abstraction layer between the tests and data. One possible implementations is shown in Figure 7.39.

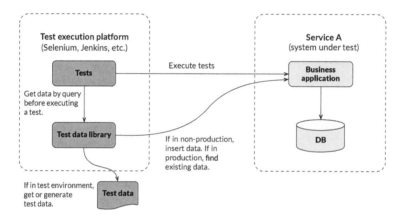

Figure 7.39. Implementation of test data generation.

In the reference implementation illustrated in Figure 7.39., irrespective of the environment, tests request the required data from the test data library. If tests are executed in the non-production environment, the test data library will generate the corresponding test data and insert it into the system under test. If tests are executed in the production environment, the test data library will attempt to find data that already exist in the system under test. In both cases, the test data library should work with the business service via its API. This API may be a regular service API or a separate CRUD API that can be used only by tests. In this case, the CRUD API should be turned off in production by a feature flag and should be protected to prohibit regular clients of the service from using it.

Although implementation of a test abstraction library and checks to ensure that any test may work with generated test data and with production data is preferred, sometimes the implementation of tests in such a way is expensive. A simpler approach would be to split all tests into two categories: tests that work with "empty" service-generated test data as needed, and tests that can work with "prefilled data" and can find the required data from the service via the API. The second type of test is generally more difficult to implement, because the required combinations of data may be missing in the service and tests should be able to work with any data that matches the pattern they search for. Under no circumstances should the second type of test use hard-coded identifiers to search for data, to avoid random test failures when the production data change slightly. Too often, in our experience, tests became so dependent on static identifiers to find required data that, after uploading new datasets to the test environment, tests begin to fail as a result of missing data and not because of actual defects. This makes tests brittle and deteriorates trust in test results very quickly.

7.4.6.3. Upgrade and Backward Compatibility Testing

Testing needs to ensure not only that the new version of the service satisfies functional and nonfunctional requirements but that both old and new versions work properly during the execution of the upgrade. For that reason, the deployment of new versions of service components to the non-production environment should use the same upgrade method as will be used in production for the deployment of that specific change. For example, if a change is planned to be deployed in production with a blue-green upgrade, it should be deployed to the test environment with a blue-green upgrade as well. If a new version of the service is backward compatible with the previous one, backward compatibility needs to be tested as well.

Let's consider an example when a service consists of a business application and database components, changes of that service are typically deployed to production with blue-green upgrades, and the service's contract needs to maintain backward compatibility. Before deployment of the new version starts in the test environment, the test environment needs to be pre-created with the version of the service that is currently deployed to production. Then, the new version of the business application deployment package is deployed to the environment in blue-green mode (Figure 7.40.).

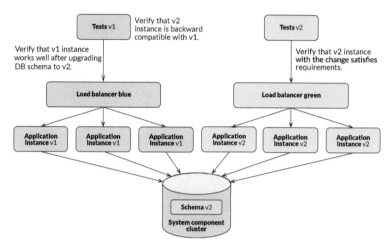

Figure 7.40. Testing of a service with a requirement for backward compatibility.

After the upgrade is executed, the following tests should be executed:

1. Tests that correspond to the old version of the service should be executed against the old (blue) instance of the service to ensure that the upgrade of database schema didn't affect the functionality of the blue instance.

2. Tests that correspond to the new version of the service should be executed against the new (green) instance of the service to verify that the new version satisfies requirements.

3. Tests that correspond to the old version of the service should be executed against the new (green) instance of the service to ensure that the new version is backward compatible with the old one.

If the testing is successful and the version is promoted to the next stage of the change management pipeline, the information about the specific version of test that was used to verify the specific new version of the service should be recorded and saved. This information will later be used to determine the specific version of test to execute against the old instance of the service.

If testing of backward compatibility is not required, the last step of testing can be omitted. If a rolling upgrade is used instead of the blue-green method, a similar approach can be used. For rolling upgrades, the upgrade process will need to be configured to pause the upgrade in the middle, when a service consists of half instances of the old version of the business application and half instances of the new version.

7.4.6.4. Dependency Management

As we discussed in the chapter about source code, we recommend storing tests and test data in a source code repository that is separate from the repository where the components' code is stored. This recommendation brings with it the issue of dependency management between tests and applications. We understand that some teams may prefer to store tests in the repository with the system under test. We recognize that this approach works as well, although it generates issues with launching a change management process for changes in tests, which are not considered actual changes. However, going forward, we will assume that tests are stored in a separate repository. In this section, we will discuss several techniques to implement dependency management between tests and systems under test to find the proper version of tests and test data to verify a certain version of the service components.

The simplest way is not to perform explicit dependency management but to always take the latest version of tests and test data to verify a new change. In a normal development process, the QA team creates and maintains tests in parallel with feature development, so the tests and test data are kept up to date. This technique works especially well when there is a regular cadence of releases, so the QA team can keep up with development. However, in this case, the latest version of the tests may not correspond to the latest version of the code and this makes the analysis of the test results more difficult.

Another approach that provides stronger guarantees is to reference a specific version of the components from the repository with the tests. In this case, the QA team needs to update this version every time they implement a new test or every time a change is implemented in any of the components. To minimize the burden of updating specific version references for every change of the component code, only major versions can be used, ignoring minor, patch, build, or hashcode of commit changes in versions. An additional policy can be implemented to automatically reject the change if the new major version of the component is higher than the version that the tests can verify. This approach will guarantee that every new feature and release is tested before it is approved for release to production.

7.4.6.5. Types of Testing

The approval of a change requires verification that the new version of the service satisfies several types of requirements. These requirements may include function, performance, stability, security, compliance, and others. Different types of requirements are oftentimes enforced by different teams, when the lead from each team needs to provide a sign off before a change can be deployed to production. This leads to the creation of different types of tests and corresponding policies. The typical types of testing, based on the types of requirements include:

1. Functional testing verifies that the service functionality satisfies the business requirements in isolation and as a part of the bigger system.

2. Performance testing verifies that the service can provide the required latency and throughput while utilizing a certain amount of system resources. Performance testing is focused on the performance of a single service instance running in a container or VM, as well as the total performance of the entire service cluster. The latter is defined as a scalability requirement. Performance testing also includes longevity and stress testing.

3. Stability testing ensures that the service functions correctly despite the occurrence of the failure of various components or other destructive actions. Obvious failures include shutting down of VMs and containers or system process, network, and load balancer failures. In some cases, stability testing and performance testing can be joined together as non-functional testing.

4. Security testing verifies that the service complies with the security requirements and standards.

From an organizational perspective, functional, non-functional, and security testing are often performed by different teams and require separate sign offs.

In addition to the different types of requirement, tests that cover the same type of requirement can be further segmented for efficiency and maintainability reasons. For example, functional tests can be split into smoke, new feature, regression, integration, and other category tests. This leads to the creation of several test suites, each of which may have special characteristics like time of execution, special requirements for the test environment, and special conditions that will mean they are triggered only for specific types of change. Typical categories of testing include:

1. The smoke test suite includes a subset of tests that verify only key requirements and can be executed quickly.

2. The service-level test suite includes all tests required to verify a certain type of requirement, but they are done at the level of an individual service. These tests can be performed with real upstream dependencies or in isolation with test data, stubs, and mocks.

3. The integration test suite ensures that a service works well with upstream and downstream integrations. To minimize the cost of creating and maintaining tests, the service-level test suite of a downstream or top-level service can used as an integration test suite for an upstream service. In many cases, a separate integration test suite is created with end-to-end tests.

We also want to distinguish several specific categories of tests for specific requirements:

1. Comparative performance testing is similar to performance testing, in the sense that it verifies the performance of the service. However, whereas true performance testing oftentimes needs to be executed on a production-like environment at full scale to measure actual performance numbers, comparative performance testing can be done on small environments. These tests do not verify absolute performance numbers but instead verify how the performance of the new version compares with the performance of the old version of the service. This is typically done on the service level.

2. Environment validation is a special kind of testing that verifies that the environment is ready for deployment of an instance of the service. Specific testing may include checking of the network connectivity, availability of external dependencies, etc.

3. Business acceptance is often performed on an end-to-end system by the business team later in the pipeline.

The full splitting of tests by requirement types and categories may lead to too many different test suites. Sometimes, such splits are reasonable and, with good tooling and team maturity, the efficiency of using only the right tests for the right changes outweighs the overheads of managing them.

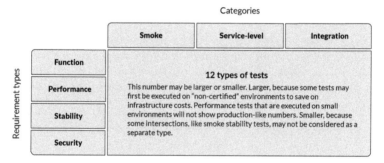

Figure 7.41. Matrix of the types of requirements and categories that need to be tested.

We understand that the N×M matrix in Figure 7.41. may not be the ideal representation of the various test types, because some test types may not be practical, such as smoke stability tests, and some may be omitted, such as comparative performance testing and business acceptance testing, but it gives some understanding of the scope of the problem that the QA team is dealing with.

7.4.6.6. Testing Pipeline

Although the testing of all requirement types is required, the level of such testing, the exact categories, and the order in which these tests are executed may be decided by each department or service team. The template for the testing pipeline for a service should be as shown in Figure 7.42.

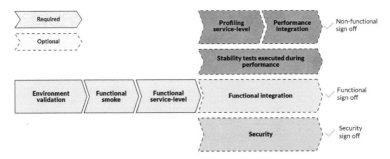

Figure 7.42. Template for the testing pipeline for a service.

Depending on the change, the execution of various test suites can be required or optional. For example, if the change is a small functionality bugfix, non-functional testing, security, and full functional integration testing can be omitted. Unfortunately, the implementation of an automated exclusion policy for certain types of tests depending on the change can be challenging, whereas manual decisions during the pipeline execution should be avoided. Therefore, the release engineering team typically implements a default pipeline that executes all necessary types of tests that are required for the majority of changes. Such a default pipeline may have the minimum required testing, and more extensive testing may be required on demand. As shown in the template in Figure 7.42., performance integration and security testing may be excluded from the default pipeline, depending on the service.

Let's review the service-level functional stages of the testing pipeline in more detail (Figure 7.43.).

Other stages of the testing pipeline that cover functional integration, profiling, performance, stability, and security tests are similar to pipeline shown in Figure 7.43. The only exception is that they may be executed in parallel and require different types of environments. One of the key aspects of each stage of the pipeline is that, after each successful execution of a certain test suite, the change should be marked appropriately. Such marking is typically implemented at the level of the source code repository through tagging of the corresponding commits, at the level of the artifact repository by labeling the corresponding deployment packages, and in a separate approvals and audit log. This marking records what approvals and sign offs each change already has. This is required to find appropriately approved changes for later stages of the testing pipeline and the change management process as a whole.

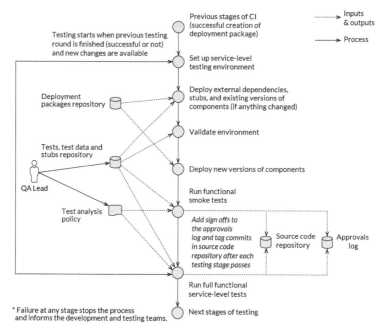

Figure 7.43. Service-level functional stages of the testing pipeline.

There are several core principles that need to be taken into account during implementation of an efficient testing pipeline. The pipeline should be implemented in a way that will:

1. Minimize the total run time of the pipeline from the beginning to the end, both in cases of failures and in cases of successful executions.

2. Identify most often defects early in the pipeline.

3. Provide sufficient information to identify the root cause of the defective component and service.

To accomplish those goals, the testing is typically broken into smoke, full-service, and integration testing, and the tests are executed in this sequence. Smoke tests run quickly and help to find the most severe defects quickly. Full-service-level testing takes more time but prevents more expensive integration testing from taking place if the service under test is defective. Integration testing is executed last, because it usually takes more time and failures of integration testing don't allow the root causes of the issues to be quickly identified.

To minimize the total execution time, some tests can be run in parallel to each other. The parallel execution of tests doesn't necessarily require separate environments for execution. For example, if the test environment is production-like, then functional, performance, stability, and security tests can be executed on the same environment without affecting each other. To accomplish this, the different types of tests should be designed in a multi-tenant way to avoid data manipulations that may affect other test suites. One exception to this recommendation is stress and destructive testing, which would specifically make the environment not operational for a period of time. Regular performance and stability testing, however, assumes that the system functionality doesn't suffer during the test, so these tests can be executed in parallel with other tests.

In many practical cases though, different environments are used for service-level testing, integration testing, and performance testing. The main reason behind this is that the environment requirements of these types of tests are different and certain savings can be achieved by having smaller environments for functional and profiling service-level testing, larger environments for integration and end-to-end testing, and production-scale environments for performance testing.

The detailed discussion of test design, test implementation approaches, and related technologies deserves a separate book. We will however put a stop here. The description provided so far should be enough for our purposes in designing the change management process.

7.4.6.7. Stability Testing

A special type of testing that becomes more important for microservices-based cloud systems is stability testing. The deployment topology of such systems is typically very dynamic, with services scaling up and down, new versions being released and deployed in blue-green or rolling modes, virtual machines or containers coming up and down with changing IP addresses, etc. The traditional approach to testing, in which specific use cases are identified and test cases are created for them, can't be easily applied to stability testing and leads to test suites that are expensive to create and maintain.

Instead, a team can implement a different approach based on randomized testing (Figure 7.44.). In this chapter, we'll give an overview of this approach; for a detailed discussion, we recommend the book *Chaos Engineering* [41].

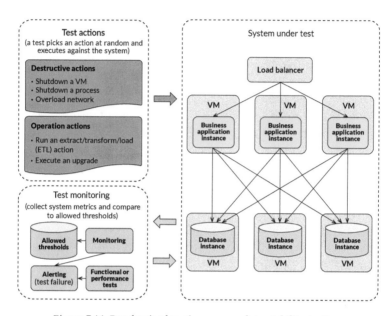

Figure 7.44. Randomized testing approach to stability testing.

The idea behind this approach is to simulate various destabilizing actions on the system under test and monitor whether the system survives these actions and continues satisfying the various functional and non-functional requirements.

Destabilizing actions can be divided into two major categories:

1. Unexpected destructive actions that are typically caused by the environment. Examples of such actions include shutdown or malfunction of a VM or container, shutdown of a system or application process, and networking issues.

2. Expected actions that are performed on the system by the production operations team. Examples of such actions include execution of an upgrade or running an ETL to upload new data.

Destabilizing actions are implemented as individual scripts. The actual test then picks an action at random and executes it against the system under test. The random execution should be timed appropriately to represent realistic scenarios and not kill the system. For example, if all of the VMs in the system under test were shut down within one minute, the system under test couldn't possibly survive. The random execution of destabilizing actions allows many different scenarios to be covered over a period of time, saves on the cost of test creation and maintenance, and allows issues to be found that may otherwise be missed by the team.

While destabilizing actions put pressure on the system under test, the monitoring component of stability testing verifies whether the system sustains the stress. This can be implemented in different ways:

1. Monitoring of various system metrics and execution of business service components' health checks. For example, after a database VM shuts down, the DB cluster and business application should still be healthy. After a while, if self-healing is implemented, the failed VM in the DB cluster should be recreated and the number of VMs in the DB cluster should be equal to three. Both the health status of the DB cluster and the total number of VMs in the cluster can be considered as invariants of the system, and the monitoring should ensure that such invariants are maintained.

2. Periodical execution of functional or performance tests. In addition to system metrics, the execution of functional and performance tests against the service is the perfect way to verify that the system functions properly.

Stability testing relies on randomized actions, so its execution is non-deterministic. To cover many scenarios, stability testing may need to be executed over a long period of time. Because of that, stability testing may be included as a part of the long-running performance testing execution. In some cases, a subset of stability tests may be executed in production.

There are tools that provide implementation of various environmentally destructive actions in cloud environments. Chaos Monkey and Simian Army from Netflix OSS are good examples of such open source tools.

7.4.6.8. Pipeline Optimization

As we discussed in the section about build and packaging, a new deployment package is created for every commit to a component's source code repository. This is done to create a direct one-to-one mapping between the original changes in the code and the deployment packages that can be deployed to production. In the case of build and packaging, this is possible because the process of creating a new deployment package takes an insignificant amount of time and the time between commits to the source code repository is typically longer than the time it takes to build a deployment package from the commit.

The situation with testing is different. The deployment of new versions of a service to a test environment and execution of even functional service-level tests may take significant time. The execution of functional integration, performance, stability, and security tests may take a very long time, up to several hours or tens of hours. During the time that one change is being verified by the testing pipeline, many more changes may be committed to the source code repository by the developers.

A typical approach to solving this problem is to execute testing pipelines in serialized fashion (Figure 7.45). In this case, the testing of changes to all components of a service is performed in a "single thread." Even if a new change is approved by the previous stage of the pipeline, testing of that change will not start until the previous round of testing finishes. In this case, a new round of testing uses the latest change that was approved by the previous stage of the pipeline. Even if there are other older untested changes, they will be ignored. The benefit of this approach is that it minimizes the cost of infrastructure resources for test environments and test executions.

The major disadvantage of this approach is that multiple changes to the source code repository are accumulated and tested together. This may sometimes lead to difficulties in separating bad changes from good ones and identifying the root causes of why the testing failed. For example, in Figure 7.45., it may not be clear whether it was change v2 or change v3 that was responsible for the failure of the service-level test. An even bigger problem is that, after identifying that one of the changes is defective, it will take developers time to fix this defect and the chances are high that the next immediate execution of the service-level testing pipeline will not include the fix and will lead to failure as well. There are three practical ways to avoid this problem:

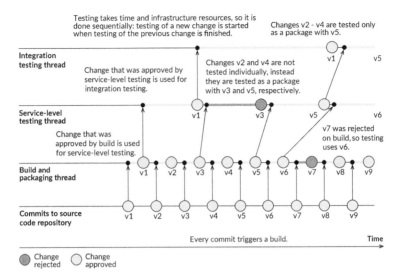

Figure 7.45. Serialized approach to executing the testing pipeline.

1. Optimize the test execution time. The longer it takes to execute tests, the more changes will be aggregated for testing and the more difficult the troubleshooting and fixing of the issue will be.

2. Let the development team execute tests on their workstations or development environments to minimize the probability of bad changes being committed to the source code repository.

3. Split big testing pipelines into several asynchronous sub-pipelines, each of which is executed in its own thread, where the shorter pipelines should identify the majority of the defects and reject most of the bad changes.

The third approach deserves a detailed clarification and a review of the typical service-level pipeline discussed earlier in this chapter (Figure 7.46).

Depending on the execution time and the probability of defects, different test suites can be executed either synchronously, when the execution of the previous test suite triggers execution of the next test suite, or asynchronously, when each test suite executes in the loop as soon as there are new changes to test. For example, based on Figure 7.46., the actual testing pipeline and execution threads should be implemented in the following way:

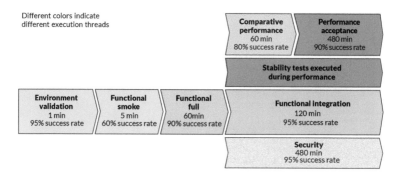

Figure 7.46. Service-level testing policy split into asynchronous sub-pipelines.

1. Functional smoke testing is executed in the loop in its own thread, because it finds most of the defects, is fast to execute, and is followed by slow test suites. As soon as it finishes, it promotes the change for functional, profiling, and security testing, but it doesn't wait for these tests to complete. Instead, it is immediately restarts to validate the next available change.

2. Functional full and functional integration testing are executed sequentially in a separate thread, because they collectively take a long time to execute and rarely find defects. The fact that they are executed in the same thread means that, after functional service-level testing finishes, functional integration testing starts. Functional service-level testing does not start for the next new change until functional integration testing finishes for the previous change.

3. Comparative performance testing is executed in a separate thread, because it finds a significant number of defects, takes a relatively short time to execute, and has only small requirements. Once it finishes, it takes the next change that passed functional smoke testing. Once a change is approved by profiling, it can be taken for verification by performance testing.

4. Performance and security testing are executed in their own threads because they take a long time to execute.

With this approach, the functional service-level testing stage of the testing pipeline that we reviewed before can be modified as shown in Figure 7.47.

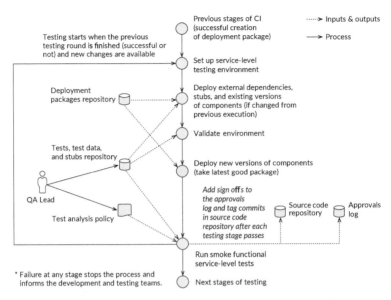

Figure 7.47. Stages of the service-level functional testing pipeline modified for asynchronous sub-pipelines.

The pipelines for other stages would look similar, except that they take the latest deployment package that was approved by the functional smoke service-level testing. Tags in the source code repository and labels on the deployment packages that are set after successful execution of functional service-level testing will allow the appropriate versions of the deployment packages to be found.

An alternative approach to solving the problem with accumulating changes is to execute the testing pipeline for each change in parallel to each other. This approach is used rarely because it leads to significantly increased infrastructure costs as a result of the creation of multiple instances of new versions of services for parallel testing.

The analysis of how many execution threads a testing pipeline for each service should have is an optimization problem that should be solved for every service. The optimization criterion is to minimize the average time to find and fix defects. This, in turn, will minimize the total average time to process each change end-to-end and will lead to the creation of an efficient change management process. The optimization should take into account the rate of commits to the source code repository, the historical or predicted probability of finding defects with each test suite, the time that it takes to execute each test suite, and the infrastructure cost that is required to execute each step. Typically, the initial pipeline is built by using predicted numbers, but it can then be adjusted on the basis of the collected statistics of execution, cost, and failure ratios. Reevaluation and reoptimization of the pipeline should be performed on a regular basis, because the statistics may change during the lifetime of the service.

7.4.6.9. Policy Inputs

When a change successfully passes the testing stage, it automatically receives a sign off from the respective test leads. To ensure legitimacy of the sign off, the test leads are responsible for providing a number of executable policies to the testing process:

1. Actual automated test scenarios, test data, and stubs.

2. Test results analysis policy and success ratio, as discussed below.

3. Information on how to find proper versions of tests, test data, stubs, and analysis policies for a certain version of the service under test.

4. Minimum test coverage threshold, if applicable.

The above policies should be provided for all types of testing, including functional, performance, security, and others. More importantly, the policies above should be automated and executable without human intervention.

Many companies are successfully implementing test automation, but oftentimes test automation includes only automation of execution of the test scenarios. Test analysis continues to be manual and is often a time-consuming activity. However, test analysis is an important part of the testing policy.

In the most straightforward cases, test analysis can be trivial and the team can establish the policy that all tests should be successful to accept the change. However, this policy is rarely achievable in practice, especially if test scenarios cover not only happy paths and core functionality but many corner cases as well. In practice, changes can be accepted and new versions can be released to production even if there is a certain number of minor defects and known issues.

To implement continuous delivery and fully automated change management, an automated policy for analyzing test results should be implemented. Most test automation tools provide test results in a machine-readable format, like XML or JSON. The test scenarios themselves can also include certain metadata about their criticality. Therefore, a set of scripts can be created to analyze test results and make a decision about whether the change can be accepted or not. For example, test scenarios can be categorized into different priorities of criticality. A test analysis script can then allow a certain percentage of failures from each criticality group of tests.

Implementation of automatic test analysis scripts doesn't eliminate human judgement. Rather, it forces testing teams to encode the policies and guiding principles that they use today in an executable policy format. This executable policy may also include clauses that, in certain cases, the decision cannot be made automatically and manual intervention is required. The goal, however, is to minimize the number of such cases.

During the testing process, all outputs and policy execution results should be recorded in a database and tied to a specific change for the audit trail.

7.4.6.10. Example: eCommerce Platform

The eCommerce platform example has a number of services (Figure 7.48), each of which should be released independently from the others. Therefore, we need to implement testing pipelines for each of them.

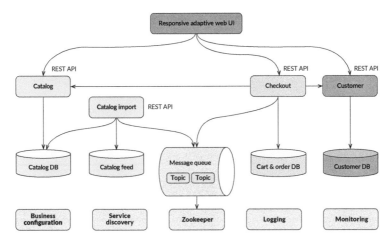

Figure 7.48. Conceptual architecture for our eCommerce platform example.

For each service, only service-level tests will be implemented, with the exception of basic component-level health checks. The test suites that will be implemented for the system are listed in Table 7.3.

Table 7.3. Test suites that will be implemented for our eCommerce platform example.

Scope	Tests
Catalog service	* Functional smoke tests using the APIs of the catalog and catalog import business application components. * Full functional service-level tests using the APIs of the catalog and catalog import business application components. * Profiling tests for the catalog service.
Checkout service	* Functional smoke tests using the API of the checkout business application component (with stubs for the catalog and customer services). * Full functional service-level tests using the API of the checkout business application component (with stubs for the catalog and customer services). * Profiling tests for the checkout service.
Customer service	* Functional smoke tests using the API of the customer business application component. * Full functional service-level tests using the API of the customer business application component.
Web UI	* Functional smoke tests using real dependencies (same as smoke integration tests). * Full functional service-level tests working with real dependencies (same as full integration tests). * Web UI security tests.
Whole system (integration)	* Functional smoke tests using the UI or the responsive adaptive web UI component. * Full functional tests using the UI or the responsive adaptive web UI component. * Performance tests using the UI or the responsive adaptive web UI component. * Stability tests.
Platform components	* Functional tests. * Stability tests.
Environment	* Environment validation tests.

In this case, the service-level tests of the web UI are equal to integration tests of the whole system. Profiling tests are implemented for the components that are most critical from a performance perspective. Stability tests are usually expensive from an execution perspective and work well on a large scale, so they are only implemented for the scope of the whole system.

Based on the different types of tests and services, the testing policy and testing pipeline for each service will be different. Let's start with the catalog service. The catalog service doesn't have any dependencies, but it has two business application components: catalog and catalog import. Both expose REST APIs, and both are covered by functional and non-functional tests. A change to any component of the catalog service should execute the same testing pipeline. In this example, let's assume that the catalog service has very well-defined and self-contained functionality and its test coverage is very good. Because of that, a change to the catalog service can be accepted if the service-level functional and profiling tests pass (Figure 7.49.). Integration testing is not required because the risk of missing a defect is extremely low and doesn't justify the longer testing times and release cycles.

Figure 7.49. Testing policy for the catalog service for our eCommerce platform example.

With the metrics above, it makes sense to split the catalog testing pipeline into two separate threads. The functional testing pipeline would then be implemented as shown in Figure 7.50.

The comparative performance testing pipeline would take deployment packages that successfully passed functional testing. Its implementation would be similar to the functional testing pipeline with the exception of tooling to implement profiling tests. Typical tools for this job include JMeter, Gatling, and Jagger.

Even if the testing policy to accept or reject a change in the catalog service is simple, integration testing for functional, performance, and stability requirements can still be performed after the change is accepted. In many cases, it may reveal non-critical defects after the change is deployed to production, so these can be fixed before they are discovered by customers.

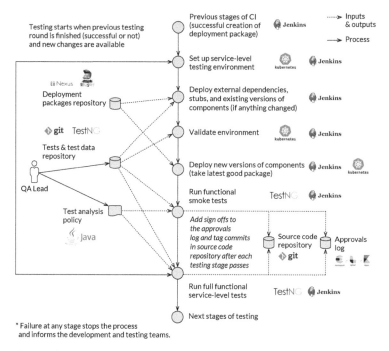

Figure 7.50. Stages of the functional testing pipeline for the catalog service for our eCommerce platform example.

The test environment for the catalog service is simple and contains only components of this service. The service is tested with generated test data. The absence of upstream dependencies means that no stubs and mocks are required.

Customer service is similar to the catalog service, with the same policy and testing pipeline.

The checkout service is different (Figure 7.51.). For our example, we assume that it contains very complex business logic. The probability and costs of defects in checkout are very high. The checkout service also contains two dependencies: catalog and customer service. To increase the quality of the checkout service and to isolate the potential defects in checkout from the defects in upstream dependencies, the service-level tests of the checkout service involve stubs for the catalog and customer services. As a result of the usage of stubs and the high cost of the risks, approval of a change in the checkout service requires both service-level and integration testing.

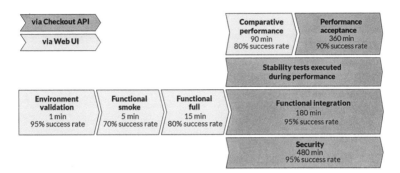

Figure 7.51. Testing policy for the checkout service for our eCommerce platform example.

The checkout testing pipeline is split into five different execution threads: functional service-level, profiling service-level, functional integration, performance integration with stability, and security. Each execution thread, or sub-pipeline, requires a different environment. The implementation of each sub-pipeline is similar to the functional pipeline implementation for the catalog service. The difference is in the environment requirements and technologies that are used to implement each type of testing:

- Functional service-level testing requires only the checkout components and stubs of upstream dependencies in the environment. The testing is done via an API, so TestNG is used.

- Profiling service-level testing has similar environment requirements but uses JMeter for test implementation.

- Functional integration requires an environment with all services and components. The testing is done via the web UI, so a Selenium cluster is required to execute the tests. TestNG can still be used as a framework.

- Performance integration requires a production-like environment with all components and services. JMeter or a proprietary performance testing tool can be used for testing.

- Security testing has similar environment requirements to functional integration testing. Specialized web UI security testing tools are used.

Tests that are implemented for the web UI can be reused as integration tests to verify the whole system. However, the critical path of the testing pipeline for the web UI itself doesn't have to include full performance and stability testing. Only functional integration and security testing is required.

A special case is the testing pipeline for platform components, such as the message queue, service registry, secret management, and logging and monitoring. Changes in these components are rare, but they may affect the stability of the whole system and defects have a very high cost. However, such changes rarely affect the functionality of the services. The default pipeline for platform components includes basic functional and health checks for changed components, as well as component-level stability testing. Major changes may also require full functional and non-functional integration testing.

As we mentioned before, the testing pipelines of all components may include critical path and non-critical path testing. Critical path testing includes the tests that absolutely must be passed before a decision can be made to accept or reject the change. The non-critical path may include long-running tests that may find defects that can be fixed after the change is deployed to production. Stability testing is a typical example of such non-critical path testing for most services.

In all of the cases above, even if the testing pipeline for some services requires integration testing with real upstream and downstream dependencies, changes in different services can still be released to production in parallel with changes in other services. This is a major assumption that requires backward compatibility of changes in the services' functionality. To ensure this, the testing pipelines of all services include verification of upgrades and backward compatibility, as described earlier in the chapter.

7.4.7. Dependent Changes

Strong service contracts, backward compatibility, and solid service-level test coverage are the goals to ensure that changes in different services can be released independently and concurrently. In some cases, however, these guarantees may not work or may be difficult to implement. One of the most common examples is when an organization doesn't have extensive experience with implementing service-oriented architecture, and the risk that changes in one service may the break backward compatibility or affect the behavior of other services is too high. Another example is when the system as a whole or groups of services are being redesigned and major cross-service changes are implemented. In these cases, changes to individual services cannot be considered independent.

In the case of dependent changes, even if the system as a whole is broken down into individual services from the architecture, development, and deployment perspectives, it is still a monolith from the change management perspective. In fact, if the service owners do not commit to strong contracts, compatibility, and change independence, an organization can't claim that a proper service-oriented architecture has been implemented. Dependent changes to different services cannot be approved individually until they have been tested in integration with other changes to other services.

At this point, we would like to emphasize that a reliance on dependent changes is not best practice. In the long term, organizations need to commit to a proper service-oriented architecture, independent changes, and the ability to release changes in different services independently. The only reason we are going to discuss dependent changes in more detail is that it may help to lower the entry barrier for some organizations, who are only just starting their journey toward true continuous delivery.

7.4.7.1. Global Version

One technique to implement dependent changes is to introduce a global version of the whole system. A global version can be implemented as a separate source code repository containing a file with versions of all services and components that are included in the system.

The file format may be as simple as a list of lines, in which each line represents a component with information about the version of its deployment package, for example, "ServiceA_DB: v1.7.523." (Figure 7.52.) The source code repository with the global version follows the same branching strategy as any other source code repository with a business application or system component. The global version file can be automatically updated with the new version of a service when that service has passed its service-level testing and is ready to be considered as a release candidate.

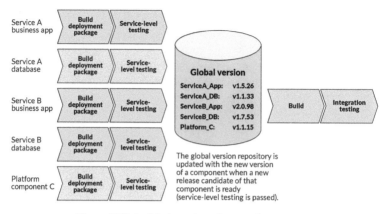

Figure 7.52. A global version of a complete system.

Because the global version represents the whole system, it can be considered as a large composite service itself. However, instead of the source code being written in a conventional programming language, the global version source code represents a composite service that consists of specific versions of other services and components, which, in turn, conceptually start to resemble libraries of traditional services. Otherwise, the global version source code is handled similarly to the source code of regular components. Changes to the global version repository trigger the pipeline, similarly to the pipeline for changes to component-level source code repositories. The global version build process is trivial and consists only of packaging the global version file, versioning it, and saving it to the artifact storage repository. Building of the global version triggers execution of the integration test suite for the whole system.

7.4.7.2. Global Deployment Package and Upgrades

Individual deployment packages are built on a component level. When multiple components in different services need to be deployed at the same time, the order in which the deployment should be executed may not be obvious. Therefore, in addition to a file with references to versions of components' deployment packages, the global version repository may include a set of scripts that are used to upgrade the whole system.

Global deployment scripts are essentially orchestration scripts that encode the order in which to execute upgrades of multiple components. Such scripts are generic enough to handle upgrades of various combinations of changed components. Depending on the exact components that need to be upgraded, the scripts may chose different orchestration upgrade strategies.

Blue-green upgrades are the typical upgrade strategy used for dependent changes across multiple components and services. Unlike the orchestration of dependent changes with rolling upgrades, blue-green upgrades allow additional instances of all changed services to be created and tested as a whole. In this case, the orchestration deployment script creates a separate "green" stack consisting of all changed components and downstream components, as described in the section about blue-green upgrades.

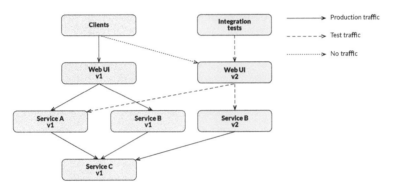

Figure 7.53. Orchestration of dependent changes implemented with blue-green upgrades.

In the example in Figure 7.53., if there are backward-incompatible changes between service B and the web UI, they can be changed together with low risk to the existing production traffic. The order of upgrades is trivial as well, and the orchestration script should know only about the direction of the dependency graph. The orchestration script then just deploys new versions of all changed services by starting from the most upstream one.

7.4.7.3. Integration Testing

The global version testing pipeline (Figure 7.54.) includes all types of integration testing: functional, performance, stability, security, etc. Integration testing is organized in the same way as service-level testing. A similar approach to the test data management is used, with the exception that stubs and mocks are rarely used. Production-like data are used when possible. Upgrades should also be tested in the same way as in service-level testing.

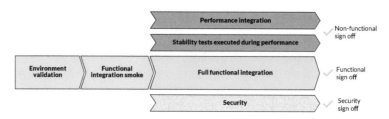

Figure 7.54. The global version testing pipeline.

Integration testing is oftentimes expensive and time consuming, so the testing pipeline of the global version is done not for every change but on a periodic basis. Nightly testing can be chosen as a compromise. The same rules apply to optimization of the global version testing pipeline as to the service-level testing pipeline.

7.4.7.4. Policy Inputs

Global version testing has the same policy inputs from the respective test leads as service-level testing. To ensure legitimacy of the sign off, the test leads should provide the following executable policies to the testing process:

1. Automated integration test scenarios, test data, stubs, and mocks.

2. Test results analysis policy and success ratio.

3. Information on how to find the proper versions of tests, test data, stubs, and analysis policies for a certain version of the system under test.

4. Minimum test coverage, if applicable.

The above policies should be provided for all types of testing, including functional, performance, security, and others.

7.4.7.5. Example: eCommerce Platform

For the purposes of the example with the eCommerce Platform, we'll assume that the organization is new to service-oriented architecture and changes to all components need an extra verification. Therefore, all changes are considered potentially dependent. As a reference, the conceptual architecture of the platform is shown in Figure 7.55.

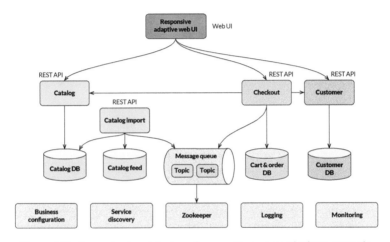

Figure 7.55. Conceptual architecture for our eCommerce platform example.

To implement the extra verification, a global version repository is created and a single text file is put in it. Each line in the file represents a component ID followed by the version of the deployment package of this component. The number of lines in the file is equal to the number of components. Both business applications and system components are included:

- Responsive adaptive web UI.
- Catalog business application.
- Catalog import business application.
- Catalog DB.
- Catalog feed.
- Checkout business application.
- Cart & order DB.
- Customer business application.

- Customer DB.
- Message queue.
- Business configuration.
- Service discovery.
- Zookeeper.
- Logging.
- Monitoring.

Every time a component passes the service-level tests, its version is automatically updated in the global version file. Components' versions can also be updated manually if a specific combination of new versions needs to be released.

Global version testing is implemented with integration tests. Web UI tests are used as functional integration tests. As we described in the service-level testing section when reviewing this example, the integration test suite also contains performance, stability, and security tests. To ensure efficiency of the pipeline, the functional integration tests are broken down into the smoke suite and full suite.

To increase the efficiency of the testing pipeline, testing is performed in three parallel threads:

1. Functional integration smoke and function integration full. Performed sequentially in the loop. As soon as the previous testing round is complete, the new round starts with the latest change in the global version file.

2. Performance and stability. Executed nightly.

3. Security. Executed nightly.

The performance and stability suite andthe security test suite are executed only on successful completion of the functional integration smoke suite. The staging of multiple changes is not a big problem for the performance and security tests because, in most cases, defects would be found in the service-level testing. If there are remaining defects, it is relatively easy to find the root cause of the defect because only a few components are not covered by the profiling tests.

When we described implementation of the service-level testing for this example, the checkout and web UI services included execution of integration tests as a part of their service-level testing pipeline. Because integration testing becomes mandatory as a part of the global version testing, the integration test suite execution can be removed from the checkout service-level testing pipeline.The execution of the performance, stability, and security test suites can also be removed from the web UI service-level testing pipeline. This improvement helps to avoid double testing and makes the corresponding service-level pipelines shorter.

7.4.8. Release

When a change successfully gets through all verification stages of the change management pipeline, it should be prepared for deployment in production. The final preparations for such deployment include the collection of the necessary release artifacts to create a change request for production. The actual decision to deploy the change to production is often manual, so the release artifacts should contain all information that is necessary to efficiently make and execute such a decision. It is worth noting that the release artifacts are collected during the execution of every step of the pipeline, not as the last step of the pipeline. At each step of the pipeline, all of the necessary information should be available for the change to be moved to the next step. With that model, production is just the final step of the pipeline. The reason we talk about release artifacts in a separate section rather than discussing them across the other sections is just an optimization to consolidate all of the relevant information together.

Release artifacts should include:

1. The set of actual deployment packages representing new versions of components, in which the change has been implemented.

2. Release notes in the form of the original business requirements that are implemented in the deployment packages.

3. Evidence of collected approvals from the business, development, testing, security, and other leads. The approvals are represented by the execution logs of the executable policies that the corresponding leads provided to the change management process.

4. Additional deployment instructions, including the method to apply upgrades for each deployment package (rolling or blue-green).

In the case of service-level releases, every release candidate will contain only the deployment packages related to that service. Multiple deployment packages for different services will exist only in the case of dependent changes and the use of a global version.

7.4.8.1. Change Lifecycle

Before we get to the discussion of specific release artifacts, we have to specify that an actual release candidate may include multiple changes. Although the goal is to have production releases as closely mapped to business changes as possible, it is not always feasible to have true one-to-one mapping. One of the reasons behind this is the asynchronous nature of the change definition, development, testing, and release pipelines. By the time the change is ready to become a release candidate, it will have gone through a number of systems and transformations.

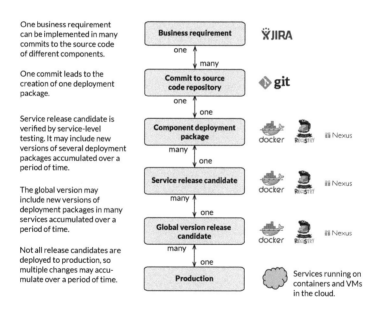

Figure 7.56 Systems and transformations that a change may go through before becoming a candidate for release.

As we show in Figure 7.56, during these transformations, changes are represented in different forms. During some transformations, an original change may be split into multiple sub-changes; during others, multiple changes may be aggregated together as bigger changes:

- After an original change is defined as a requirement in Jira, it may be implemented either in the source code for several component either all at once or over a period of time. In the former case, a change is broken down into multiple smaller changes in "space;" in the latter case, it is broken down into multiple smaller changes in "time."

- During the next step, the change is represented as a commit to the source code repository of a single component. Each commit to a source code repository leads to the creation of a single deployment package, which is stored in an artifact repository.

- In the next several stages, the change is represented by a new version of a deployment package that contains the full information about the corresponding component, including its executable binaries, also known as deployable units and deployment scripts. However, because the minimum release unit is a service, it may contain multiple new versions of its components' deployment packages. Furthermore, as we discussed in the chapter about service-level testing, the testing may not be executed for every creation of a new deployment package. This leads to multiple changes being accumulated as new versions of deployment packages before the latest version is validated, approved, and becomes a release candidate.

- If changes in the system are dependent, the global version will further aggregate multiple changes across multiple services and as a result of long-running integration tests. Conceptually, this stage is similar to the previous stage of aggregating multiple components or libraries in a single composite service.

- Finally, when a release candidate is deployed to production, the actual change is now represented as a new state of the production environment in the form of new versions of services running on containers and VMs in the environment. As with testing, not every release candidate will be deployed to production, which leads to an accumulation of changes.

In all cases, the original changes may be split or aggregated in both "space" and "time." The more changes are split and aggregated, the harder it is to reason about an individual change and the more error-prone, risky, and uncontrolled the change management process becomes. A well-designed architecture and continuous delivery best practices minimize such splits and aggregations. The architecture helps to minimize splits and aggregations in space, because the business changes are more easily mapped to individual services and components. Well-designed service contracts help to avoid the use of global versions in the long run. Continuous delivery practices, including automation and executable policies, minimize aggregations of changes over time by releasing changes once they are ready.

Nevertheless, even if all best practices are implemented, there will be cases when the relationship between the final release candidates and the original business requirements will not be one-to-one. This raises the importance of properly tracking changes during the transformations in the change management pipeline.

7.4.8.2. Tracing Back to Business Requirements

When a release candidate is prepared, one of the goals is to find information about what original business requirements are implemented in the set of deployment packages that are part of the release candidate. To automatically collect this information, the original business requirements should be traced back from the release candidate on the basis of the change lifecycle. Let's review how this tracking can be implemented.

Irrespective of whether the release is performed on the service or global version level, it contains versioned deployment packages of all components. We will review the case with the global version, because it is more complex. In this case, the release contains a specific version of the global version file, which, it turn, references specific versions of components' deployment packages. Unfortunately, the specific versions of the deployment packages will not provide sufficient information about all of the changes that are included in the release, because multiple changes may have been accumulated over time since the last production deployment. To find all of the changes that this release includes, we need to calculate the difference between the deployment package versions of this release candidate and the versions of deployment packages that are currently deployed in production.

The information about the current versions of the packages deployed to production can be found in the system responsible for tracking the status of changes or in the production configuration management database. The implementation options for such a system will be discussed later in the section about change tracking. One of the implementation methods is to use tagging of corresponding commits to source code repositories. For example, each time a commit to the global version is successfully tested or released to production, a corresponding tag is created on the appropriate commit. Information about the global version that is currently deployed to production can then be obtained from the global version source code repository. The same technique can be applied to individual component-level source code repositories.

The next step is to find the commits to the source code repositories on the basis of the deployment package versions. The relationship between specific deployment package versions and commits to the corresponding component-level source code repositories is one-to-one, so this task is trivial. The most typical approach is to include a commit identifier in the deployment package.

Once the commits of existing production versions and release candidate versions are found for each component, all intermediate commits with commit messages can be extracted from the source code repositories. In order to trace the commits to business changes, a rule should be enforced at the source code repository level that every commit should reference an identifier of a business requirement (story, task, or feature) that is implemented as a part of this commit. This is a common best practice that can be enforced by many popular source code and requirements tracking tools. Once this rule is implemented and every commit references a business requirement by ID, the requirements can be parsed from the commit messages.

At this time, the only challenge left is to separate business requirements that are fully implemented in the new version of the deployment package from those that are partially implemented. Because business requirements may be implemented and adjusted in many commits over time, there should be additional information about the level of completeness. This information may either be provided as a part of the commit message or extracted from the requirements management tool by using the IDs of the requirements.

Some pre-integrated tool suites available on the market provide most of the functionality described above out of the box. In other cases, simple custom functionality can be implemented to enhance the tooling. In any case, the above process allows information about implemented business requirements to be collected for each release candidate in an automated fashion. This information becomes a cornerstone of the release notes that should be included with the release candidate. To make this approach work, any change should be well annotated, both at the code level and at the task, feature, story, or business requirement tracking level. For example, commits referencing one-line requirements in Jira or simple "improve performance" commit messages should be rejected.

7.4.8.3. Approvals and Audit Log

To allow an educated decision to be made quickly about whether the release candidate can be deployed to production, the full list of approvals should be collected and included with the release candidate. At every step of the change management pipeline, the results of the step should be recorded and stored in a database or log together with a unique identifier for the change at that stage.

The implementation of such approvals and the audit log can be relatively straightforward. In the simplest form, such a log can be implemented with the existing functionality of the tools that are used to manage requirements, perform code review, build source code into deployment packages, and execute tests. As long as those tools are pre-integrated and log retention can be configured in each tool, the collection of all approvals and presentation of them in a convenient form is trivial.

In other cases, additional custom implementation may be required to store the log and extract the entire chain of approvals per change.

7.4.8.4. Immutability

One non-trivial requirement that must be ensured for the release candidate is that, after the release candidate is created, it cannot be modified. This specifically means that no deployment packages included in the release candidate can be modified or tampered with. Although, conceptually, this requirement may seem simple, its practical implementation with the tools popular on the market may be challenging.

If the change is only referenced in the approvals and audit log by its version, deployment package filename, or tag in the source code repository, the content of the change may be modified without changing of the identifier. For example, Git may allow a tag to be moved to a different commit, and Sonatype Nexus may allow rewriting of a file under the same name.

The implementation of immutability of changes at all stages of their lifecycle requires additional measures. First, proper configuration of access control in the tools will reduce the most obvious risks. Second, although traditional versions may still be used for human readability, the hash sum of a change's content should be added to the approvals and audit log at every stage. For example, every time that a new version of a component is deployed to a test environment, the audit log should include information about the name of the deployment package, its version, and hash sum of its content. The audit log of the testing stage should also include the hash sum of the content of the tested deployment packages. When that deployment package is used for production deployment, after it is extracted from the storage, the hash sum of its content should be calculated and compared to the hash sum that was logged during testing.

Some out-of-the-box tools that manage business requirements, code, and artifacts may provide mechanisms to ensure the uniqueness of changes and prohibit modification of changes without special records in internal audit logs. In other cases, custom controls may need to be implemented. It goes without saying that whatever technology is chosen for the approvals and audit log, it should ensure immutability of the log and protection from modifications.

7.4.8.5. Example: eCommerce Platform

As we discussed previously, the release candidates of the eCommerce platform are based on the global version and contain the usual artifacts:

1. Deployment packages of the changed components, as well as the deployment package of the global version containing the orchestration deployment script.

2. Release notes, including the original business requirements implemented in the packages and approvals log for the global version, as well as the individual components. The approvals log includes sign offs from the business, development, functional testing, non-functional testing, and security leads. Sign

offs are based on automatic execution of policies provided by the leads during the change management pipeline.

3. Upgrade method to use for each component, if different from the default. By default, blue-green upgrades are used for all components apart from system-of-record databases.

The approvals and audit log is available in Jenkins, because it manages the change management pipeline. Jenkins is, in turn, integrated with Git, SonarQube, JMeter, and TestNG to "see" the approvals or rejections of changes at every stage of the pipeline. The change tracking throughout the lifecycle occurs through the integration of Jenkins, Nexus, Docker Registry, Git, and JIRA. To connect the changes between the source code and deployment packages, the commit hash codes from Git are added to each deployment package.

In addition to Jenkins, the audit and approvals log can be implemented by using an ELK stack (comprising Logstash, ElasticSearch, and Kibana). This allows better control over its modification and retention policy. This stack also helps in finding approvals for proper changes later and in collecting metrics and statistics for the pipeline execution stages.

To ensure immutability, the access to Nexus and Docker Registry is limited and the changes use the hash sum of content as a unique ID, in addition to a human-readable version.

7.4.9. Production Deployment

When a release candidate is ready, the final decision is made on whether to deploy the release candidate in production. The decision may require manual approvals from release engineers and the change management board, but because all approvals and sign offs have already been collected and the appropriate evidence included in the release artifacts, the decision should be very efficient and almost automatic. In the case of manual approvals, the necessary sign offs may be collected in JIRA and ServiceNow, where the formal change requests are created. If decisions are automatic, existing approvals collected during the CICD pipeline may be sufficient. It is in everyone's best interest to deploy release candidates in production every time they are ready. Frequent releases minimize the risks of breaking changes by minimizing the number and size of the changes that are deployed to production at the same time.

At this point we should clarify that production deployment is a complex process. In order for the change to be visible to the end customers, multiple steps may need to happen:

1. Deployment packages with the change should be deployed to the production environment.

2. Production traffic needs to be switched to the new instances of components and services.

3. If the change is protected with feature flags, those feature flags need to be enabled.

Steps 1 and 2 are typically included in the upgrade procedure, because the final state of either rolling or blue-green upgrades is when the production traffic goes to the new version of a service. For simple changes, both steps can be executed without delay and the third step may be omitted altogether.

However, for significant changes, the actual switching of the traffic in production from the old version to the new one may take significant time, especially if canary releases or A/B testing are being used. The transition between each of the steps above may require additional testing and approvals, which we will review in this section.

We will use the upgrade of a top-level business application of a single service for this example. The handling of upgrades deep within the service graph was described earlier in the section dedicated to deployment.

7.4.9.1. Starting the Upgrade

The deployment to production starts with a new version of the deployable unit being deployed to production without the old version being destroyed yet (Figure 5.57). In the case of a rolling upgrade, the new version will automatically start accepting traffic right away and, technically, the process will go into the canary release stage. In the case of a blue-green upgrade, the new version is deployed alongside the previous one.

The new version may be deployed at a small scale to save resources. After the new version is deployed, it is verified with non-destructive smoke tests. The tests should be able to work with the production data without destroying or modifying it. This may be accomplished by creating test data that don't interfere with the production data. Alternatively, the tests may use queries to find the proper data from the actual production dataset.

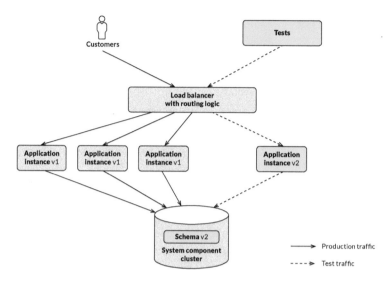

Figure 7.57. *In the first step of production deployment, the new version is deployed alongside the old one.*

7.4.9.2. Acceptance Testing

If acceptance testing and final approval by the business is required, the best time to do it is after deployment but before switching traffic to the new version. The new version may be exposed on a different endpoint that is available only to a number of internal users (Figure 7.58.).

After the business stakeholders see the new feature and approve it, traffic may be switched to the new version.

7.4.9.3. Canary Releases and A/B Testing

Exposure of the new functionality to the customers may be implemented in several ways and done in several stages. The first stage is switching a portion of or the full production traffic to the new version. If a portion of the production traffic is switched first, this stage is called a canary release (Figure 7.59.).

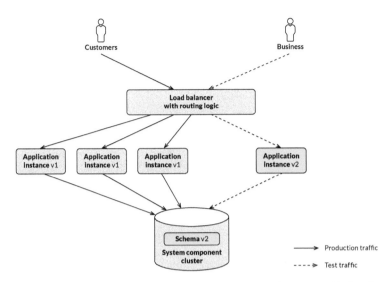

Figure 7.58. The new version is exposed to the business for approval before it takes any traffic.

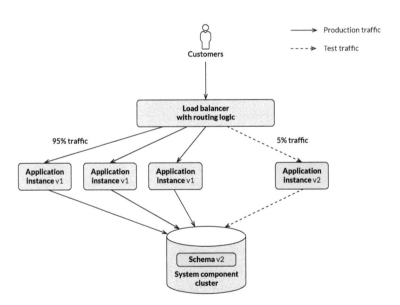

Figure 7.59. The new version is exposed to a portion of the production traffic in a canary release.

The selection of customers to route to a particular version can be done at random or via a separate configuration according to certain customer attributes. The routing is implemented on a load balancer level, which typically has routing capabilities. In the case of customers interacting with the system over a web UI, the actual routing decision is done with a setting cookie, which is used to understand where to direct a request from the customer. In the case of upgrades deep within the services graph, the routing can be done either at random or by passing cookies from top-level web UI components. The latter requires implementation of the routing fabric that will be discussed in the Further Improvements chapter later in the book.

If the canary release is successful, the traffic can be fully switched to the new version. The exact policy of the switch depends on the change and on how fast a feature should be made available to the entire customer base. In some cases, two separate instances of different versions of services can be maintained for A/B testing (Figure 7.60.).

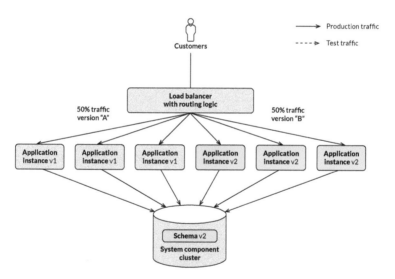

Figure 7.60. Both new and old versions are maintained to allow A/B testing.

If testing of the new feature is successful, in the end, all customers are routed to the new service instance and the old one is destroyed.

Canary releases can also be implemented with rolling upgrades. However, in this case, the load balancer will not be able to control what versions receive the traffic from specific customers. Therefore, the requests will be routed between the new and old versions at random. The exact percentage of traffic that the new version will receive will depend on the percentage of application instances of the new version, not on the configuration of the load balancer.

The behavior described above can also be implemented by using feature flags. In this case, the switching of traffic is a technical procedure and can be performed quickly. By default, the new change is turned off with the corresponding feature flag. Business users can then enable the feature for a part of the entire customer base by using a business configuration tool. This is described in more detail in the section about configuration changes.

The use of feature flags to enable new functionality in production is especially useful when rolling upgrades are used for deployment, because rolling upgrades are fast, but can't control the routing of specific requests to specific versions. Feature flags may also be used with blue-green upgrades, but in this case, there is some degree of redundancy. For obvious reasons, the use of feature flags for canary releases and A/B testing puts more pressure on the application development team, whereas the use of blue-green upgrades and load-balancer-level routing requires more effort from the deployment engineers and operations team. Both approaches have some pros and cons. Infrastructure-level blue-green upgrades provide higher isolation between versions, more control, and less risk. Application-level feature flags, on the other hand, provide higher flexibility and simpler and faster upgrades. In both cases, the implementation is part of the code, either the application code or the deployment code, and the configuration changes can be allowed via business configuration dashboards. The choice is, therefore, a matter of taste and the skills that are available in the organization in different teams.

The actual implementation of a change in production happens only after a substantial percentage of customers start using it. The deployment of new packages to production, acceptance testing, and canary releases can be considered as the final stages of the change management process.

7.4.9.4. Example: eCommerce Platform

The deployment of the eCommerce platform was described in detail in the corresponding chapter earlier in the book. Let's review the production deployment procedure by using a dependent change in the catalog and checkout services' business application components as an example. We will use blue-green upgrades with required acceptance testing and canary release as an example. Feature flags will be discussed in detail later in the book in the section about configuration changes, so we assume that they will not be used for this example.

The first step of the deployment is to deploy new "green" instances of the business application components and the web UI component (Figure 7.61.). The web UI component needs to be deployed, because it is a downstream dependency of both services. The new instance of the web UI is needed for end-to-end testing of the new features and to enable the canary release for a subset of customers.

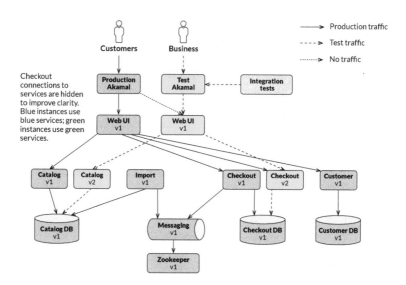

Figure 7.61. Deployment of new instances of components in our eCommerce platform example.

After the integration tests have been finished successfully on the test Akamai endpoint, business acceptance testing can start. When this is done, the production Akamai endpoint can be reconfigured to select 1% of customers and route them to the new web UI endpoint. When, after some time, it is clear that the new version is working well, all customers can be switched to the new versions over a period of time.

If the new versions of the business applications are initially deployed at a small scale, they should be scaled up to full scale before they start serving full customer traffic. If auto-scaling is used, customer traffic should be switched gradually to allow the auto-scaling to keep up with the new traffic. However, it is better not to rely fully on auto-scaling but to proactively scale new instances to the same scale at which the old instances are running after the canary stage of the release is finished.

When the old versions stop serving production traffic, they should be scaled back and then removed completely from production. If the change is big and risky, it may be beneficial to leave the old versions in production for some time for a quick and easy rollback. In this case, rollback is as simple as configuring the production Akamai endpoint to reroute traffic to the old version; no additional deployments are required.

If a routing fabric is implemented, the upgrades of services deep in the services graph can be performed more efficiently. The upgrade above could then be performed with a single Akamai endpoint and without creating a separate web UI instance. We'll discuss routing fabrics in the chapter about further improvements later in the book.

7.4.10. Tracking Changes

When changes go through the stages of the CICD pipeline and receive approvals or are rejected, their status needs to be tracked appropriately. The tracking of a change's status is required for several reasons:

1. Management and various team members need access to up-to-date information about what changes are at what stages of the pipeline and when a certain new feature will be ready. This information should be available via reports and real-time dashboards.

2. Information about he component versions that have passed various stages of the pipeline needs to be available for the proper configuration of integration test environments. For example, if a new version of service A needs to be tested in integration with service B, the appropriate version of service B should be used. In most cases, it should be the version of service B that is currently deployed in production, but sometimes

it may be the version of service B that is the latest release candidate that has received all approvals but has not necessarily gone into production. The piece of the pipeline responsible for the preparation of the test environment for service A needs to find the information about the proper version of service B in a machine-readable format so that it can deploy it to the appropriate test environment.

3. When release notes for a release are being gathered by automatic script, the script needs to find the versions of the components currently deployed to production to be able to calculate deltas between new release candidate versions and current production versions.

4. The production operations team and management need to know the current state of the versions of components deployed into the production environment.

The list of reasons could be continued, but the key point is that the information about the current status of changes should be available in both a machine-readable format and a human-friendly representation.

One of the challenges with the implementation of a good change tracking tool is that a single change is represented in different ways and stored in different tools across its lifecycle. As we showed before, the most common incarnations of the change include a business requirement (which may, in itself, be recorded in different tools during identification, clarification, and other stages), a commit to the source code, a set of artifacts representing a component deployment package, optionally an aggregation of deployment packages in the global version source code commit, together with a global version deployment package, and finally a running instance of a component in the production environment. What makes things difficult is that all of the representations of the change are tracked in different tools, and at different stages of a change's lifecycle, some of these representations may not yet be available.

Figure 7.62. shows a set of representations of a single change (without a global version) and the coarse-grained stages of approvals and sign offs that the change goes through during the change management pipeline. At every point in time, information about the status of the change should be available and easily retrievable by automation scripts and team members. For example:

- When a change is developed, the status of the original requirement in the business requirements management tool should be changed appropriately.

- When a change passes through various stages of testing, the deployment packages in the artifact repository, commits in the source code management system, and original user stories in the business requirements tracking tool need to be labeled or tagged appropriately to represent the latest status of that change.

- The original change request in the form of the user story should not be fully closed until the components containing the implementation of that user story are deployed to production.

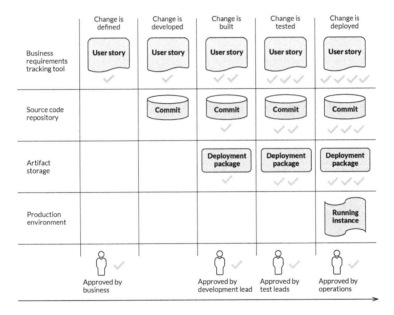

Figure 7.62. Representations of a single change during the change management pipeline.

Change tracking may be implemented with a number of tools. First of all, changes are already stored in a number of tools, like the business requirements management tools, source code management system, and artifact repository. These tools should be used to track the status of changes appropriately. Specifically:

- Business requirements management tools, like JIRA, already support various statuses of requirements and pre-defined or custom workflows for those requirements. This functionality can be used to track changes not only during development but also when they are going through the stages of the change management pipeline.

- Source code management tools, like the various implementations of Git, support the tagging of commits. This tagging can be used to track the status of specific commits when they pass through the later stages of the change management pipeline.

- Artifact repositories, like Sonatype Nexus or Docker Registry, support labeling artifacts. Therefore, the deployment packages stored in these repositories can be labeled when the packages are signed off by the test leads or deployed to production.

- Information about the versions currently deployed to production can be fetched from the production environment itself or from the tools that are used to deploy packages to production. Systems like Kubernetes support this functionality in part. However, in some cases, separate production configuration management databases may be used.

The orchestration of the status updates for a change in all systems should be performed by the tool used to implement the CICD pipeline. Jenkins would be a perfect example of such a tool.

If a proprietary pre-integrated set of tools is used, like the Atlassian suite, Microsoft suite, or Thoughtworks suite, updates to a change's status in different tools may already be implemented out of the box. Unfortunately, even these suites do not include truly end-to-end management of changes and certain aspects may be lacking, typically when it comes to the production configuration and late stages of the CICD pipeline. Open source suites also come with a set of integrations. For example, Jenkins comes with integrations with Git, JIRA, and Nexus. In all cases, some level of customization will be required.

We feel that one piece that is typically missing in most of the tools is a decent dashboard that will show robust and customizable reports on where each change currently is, what changes are currently at the same stage, etc. This is especially true when such reports need to be implemented per organization, rather than per component or per service, and when true end-to-end tracking is required from the definition of changes to production deployment. Proprietary pre-integrated tool suites do a better job at providing some dashboards; however, open source tools can provide better opportunities for customization.

In some cases, it may be a good idea to implement custom dashboards. In addition to tracking of the change status in appropriate tools, a separate approvals and audit log needs to be implemented in a separate system. The log contains all of the information about all changes, as well as the statuses of changes and times of transitions, so the dashboards may be created on top of that log. If technologies like an ELK stack are used, dashboards and reports may be a core part of the suite and the implementation of a proper dashboard may require only configuration work and less custom scripting.

7.4.11. Changes and Time

A perfectly efficient change management process should allow the implementation of any number of any changes in parallel and instantaneously. This perfect scalability and performance would allow all changes to be completely isolated and independent from each other with the verification and deployment process taking no time. Unfortunately, even with the ideal architecture, deployment, and testing automation, this is not possible. As soon as different changes affect the same component, with the source code located in one source code repository, those changes will not be independent. As soon as any testing and deployment is required, they will take time. This introduces a set of practical challenges:

1. As soon as a developer finishes working on a change, he will switch context and start working on another change. If the previous change is rejected later in the pipeline, the developer will need to return to the old change and will potentially have to stop work on the new change.

2. If a development team finish several changes during a day and then discover that one of the previous changes was rejected by testing, the fix of the defect will be committed in a repository containing a number of new changes. Instead of retesting only the fix to the old change and releasing it, the new testing pipeline will test all of the changes, which leads to an increased risk of the pipeline failing again.

Both challenges lead to delays in the change management process, accumulation of changes, and inherent risks, as well as general inefficiencies for the development teams.

The good news is that development and release management teams have plenty of tools to help with the aforementioned challenges. First of all, a microservices architecture, or at least a good service-oriented architecture, allows the system to be split into services, in which the changes for each can be isolated and completely independent from each other. Services, in turn, are split into several components, each of which is stored in its own source code repository and produces independently deployed artifacts.

Within a single component, feature flags and feature branches can be used to isolate changes. None of these tools provide perfect isolation, but as long as the team is aware of the purpose for these tools, each tool can be used with a good degree of success when applied for the right use case.

Automation of all stages of the pipeline, shortening of the pipeline, and organization of different stages of the pipeline so that the early stages are fastest and find the most defects all help to fix the majority of defects early in the process. A reduction in the number of environments and an organization to bring development and testing as close to production as possible should also help to optimize the process by reducing the time that it takes to develop and release each change. As long as the change management process satisfies the fundamental policies, the process itself should always be refactored and optimized to become as short as possible and to bring the developers as close to production as possible. We will discuss some of these ideas in the chapter about further improvements.

We hope that, at some point in time, the problem of change management in software engineering will be formalized and solved mathematically or algorithmically. The inputs to the total problem would include:

- System architecture and breakdown of the system into services and components.
- Typical changes and what services and components the changes typically affect.
- Size and productivity of development team.
- Different test suites with their metrics, including typical execution time, success ratio or percentage of executions that find defects, and infrastructure cost of execution.

The optimization goals are to minimize the average time from when the change is developed until the time when the change is deployed to production, as well as to maximize the number of changes that can be processed from commit to production within a period of time. The optimization goals may be defined per service or per system.

Given enough historical data, it should be possible to suggest improvements in how the pipeline should be organized and how the system should be split into components. It may also be possible to find out where the organization's resources should be invested and what pieces of pipelines for what services should be improved first. Of course, the actual implementation of fast automated tests and test analysis, as well as the proper use of feature flags and branches, will require work from the development team.

7.5. CONFIGURATION CHANGES

Separation of the application code from the application configuration is an optimization mechanism from the change management perspective. It enables the implementation of certain types of change without rebuilding of application binaries and without execution of the full CICD pipeline. In addition, it should be possible to apply configuration changes to production without application instances being restarted. The configuration change management process is extremely lightweight, and many of the changes are executed directly in production without any testing.

The configuration changes that we will discuss in this chapter are related only to the environment configuration. As we discussed earlier in the book, application configuration changes are managed in the same way as regular code changes. The environment configuration, on the other hand, is more conceptually similar to customer data changes, so in fact, changes to the environment configuration may not be even considered as changes because the application was pre-tested to work with any data and environment configuration within certain boundaries. Like when a customer updates their contact information through the website, the updating of pre-tested feature flag settings by a product manager or scaling up of the application instance by a support engineer may not be considered a change. The only reason that a lightweight change management process is required for configuration changes is that these changes affect the entire system, whereas changes to customer data by a customer affect only that particular customer.

The separation between environment and application configurations is logical. On the physical level, the configuration may be supplied to an application via property files, centralized services with underlying configuration databases, or other means. Even if a piece of configuration is managed as a property in the application property file, that piece of configuration may still logically belong to the environment interface. Typical examples of configuration changes include:

1. Feature flags and the business configuration of an application.

2. Endpoints to upstream dependencies.

3. Scalability configuration.

4. Load balancing and traffic routing rules.

The motivation behind configuration changes is clear. Unfortunately, the ease of implementation of run-time changes in configuration often doesn't correlate with the actual risk that such configuration changes pose. The fact that certain application functionality is moved from the application code to the application properties doesn't mean that these changes are actually low risk and lightweight. The difference between building an application code into binaries or changing properties in the run time is not what should really determine the separation between service changes and configuration changes. In fact, there are many scripting programming languages, like Javascript, PHP, and Python, for which the build is not required and, technically, the application source code files can be modified in production in the same way that application properties can be modified. However, this doesn't mean that all changes in applications implemented with these technologies are configuration changes. In the same way, some configuration that is exposed via application properties shouldn't be treated as configuration from the change management perspective.

One of the factors to consider in deciding whether a change should be a configuration change is the risk factor of that particular change. Even if a high-risk change can be implemented directly in production without building a new deployment package, it should be prohibited under the normal change management process. We started covering this topic when we discussed application and environment properties in the chapter about service deployment. In this chapter, we will review all cases of configuration change in more detail.

7.5.1. Configuration Boundaries

In order to be classified as configuration, a change should affect the application behavior within well-defined boundaries. Let's review these boundaries with the examples of feature flags and scalability parameters.

In general, the modification of feature flags and the relevant business configuration may significantly change the application behavior, so these changes have a high risk. In order to consider them as configuration changes and allow their implementation in production with a lightweight change management process, various combinations of feature flags need to be pre-tested during the service change management pipeline. The different configurations that were pre-tested then define the boundaries in which business users may change the configuration with a lightweight process and without retesting. For example, let's assume a service has feature flags A, B, and C, and all of them have only two states: "on" and "off." Let's say that during testing, the combinations shown in Table 7.4. were verified.

Table 7.4. Combinations of feature flags in our example situation.

Feature flag A	Feature flag B	Feature flag C
on	off	off
on	on	off
off	off	off
off	on	on

These combinations then define the boundaries of the configuration changes. If a change transitions the service from the A on, B off, C off state (row 1) to the A on, B on, C off state (row 2), then it is low risk and can be considered as a configuration change. If a change tries to enable all feature flags (absent from table), then the change is high risk and should not be allowed.

In practice, this policy may not be followed precisely because some feature flags may be considered independent and some feature flags may be considered low risk. Therefore, only a limited set of combinations of feature flags may need to be tested, which thus reduces the testing time in the service changes pipeline. In both cases, however, the risk is relatively high, because human judgement is required and this judgement may be imperfect.

In general, the approach of testing various combinations of feature flags is not different from testing different customer use cases with different customer data patterns. Feature flags and the business configuration should be treated in the same way as customer data.

Configuration boundaries are not limited to feature flags. In the case of horizontal scalability, the number of individual instances of an application is a configuration parameter. Even for embarrassingly parallel applications, scalability boundaries should be tested. For example, an application may scale well between 3 and 100 instances, but anything below or above these figures may lead to stability or performance problems for various reasons. In this case, such boundaries should be found during the performance, stress, and stability testing and should be documented. Hence, a request to scale this application to 150 instances in production should not be treated as a configuration change and should be rejected. This change request may still be valid, but it may require implementation of changes in the application code and application deployment scripts or changes in underlying system components.

As with feature flags, the scalability and other non-functional configuration parameters should be treated in the same way as customer data, and proper boundaries should be found and verified during testing. Ideally, the application itself should implement certain guardrails prohibiting changes in configuration that have not not been tested and that are outside the boundaries, just as an application would prohibit the modification of customer data outside the allowed boundaries, like email addresses with the wrong format.

7.5.2. Feature Flags

Business users should be able to conFigure application behavior within reasonable boundaries without going through the full CICD pipeline of service changes (Figure 7.63.). We assume that business users are not technology savvy, so they require a convenient portal to do this. Business users rarely care about the underlying architecture of the system, so the portal should cover the business configuration of all services within a business domain. As soon as a business user changes the configuration in the portal, the appropriate services need to be notified about the change and adjust their functionality appropriately. As we are talking about the business configuration, the system components would rarely be affected, and the portal will work only for the business application components of services. The integration between the portal and the application components may be done via any interface or protocol, including files, REST, or proprietary protocols of the tools used as portals and databases for the business configuration.

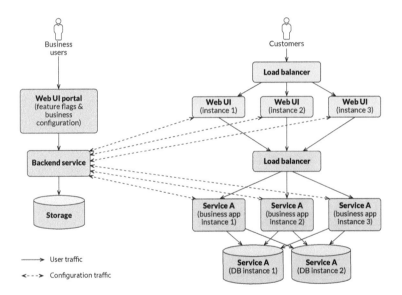

Figure 7.63. Setup that allows changes to the business configuration of an application.

In order to implement centralized management of the feature flags and business configuration, a separate service should be implemented. The service can be either fully custom built or can rely on existing tools like Zookeeper + Zookeeper UI, SpringCloud Config Server, Consul, or proprietary tools, many of which are available in SaaS mode. In any case, the service should provide:

1. A dashboard for business users where they can see the current configuration and edit the configuration as needed. The dashboard should implement proper security controls with authentication and authorization to prevent unauthorized changes. It should also be possible to implement the request and approval mechanism in the dashboard so that configuration changes pass through peer review.

2. The backend service and storage that stores the configuration, allows applications to read it, notifies registered applications of changes to their feature flags and configuration items, and aggregates the information from application instances about their actual configuration.

3. An audit log to ensure that all changes and their initiators are recorded.

Although open source tools provide a solid backbone for implementing the functionality above, in our experience, none of them provide full end-to-end implementation out of the box.

For example, Zookeeper provides a good backend storage and service for business applications to read configuration and register for notifications of changes. There is a Zookeeper UI component that can be put on top of Zookeeper to provide a basic web user interface for business users. Basic authentication and authorizations are supported as well. However, custom implementation will be required for the audit log and peer review with request and approval chains, as well as up-to-date collection of the current actual configuration from individual application instances. However, this custom implementation may be reasonably lightweight as well.

The audit log can be implemented by using existing production logging tooling. The request and approvals chain can be implemented either in the custom code of the portal or an approach that is used in the SpringCloud Config Server can be reused. In this case, a system like Git can be used for backend storage, and the typical Git peer review functionality can be used for change requests and change approvals.

The issue with the collection of up-to-date actual configurations from application instances requires separate clarification. Most of the open source tools allow application instances to read and be notified about configuration changes. However, at the time of writing this book, we were not aware of tools that allowed application instances to report back on whether they were able to actually apply the configuration parameters. This information is crucial during root cause analysis if something goes wrong. To implement this verification, custom code needs to be developed. A typical approach is to implement special REST APIs or web pages where application instances can report their current configuration. This information can then be aggregated on a centralized portal. Alternatively, business users and production operations engineers will be able to access individual instances of applications and view the up-to-date and actual information about the current configuration.

Conceptually, changes to the business configuration should follow similar high-level policies as service changes. For example, only authorized changes can be implemented in production and changes should be tested before on an environment that is indistinguishable from production before being applied to production. However, the process for implementation of those policies can be made extremely lightweight. The reason why they can be made lightweight is that all of the actual testing was performed during the service testing pipelines, when the appropriate features were implemented and verified in the business application component code. From an application perspective, feature flags and business configuration changes are nothing but changes in data.

The last challenge with feature flags that we will discuss is how to use feature flags in a microservices architecture. In most cases, feature flags will control isolated features in respective services and they could be turned on and off independently. However, sometimes, business features may span multiple microservices and will need to be turned on and off in one transaction. The straightforward approach to enable a feature in this case is to let business users turn on several feature flags separately in parallel, with each feature flag corresponding to a single service. This approach has issues because the enabling of features is an asynchronous and eventually consistent process, which means that features may be on and off across services during the period of eventual consistency. Depending on the implementation of services, this may cause problems. Another approach is to conFigure distributed features on the top-level applications and let the stream applications pass the feature flags in the request metadata to downstream services. This may require implementation of additional code in each service, but it will ensure consistent rollout of features across services.

7.5.3. Dependencies

Changes of configuration related to external dependencies should always be considered as configuration changes, because they belong to the environment configuration and not to the application itself. Among the most common examples of dependencies are the endpoints and authentication information of upstream services, which we will focus on in this section. As long as the endpoints are valid and point to the right services and the dependent services are working properly, the application should be able to use any value. In fact, the process of configuring application instances with the correct endpoints should be automated, and this information should be exposed by the environment as one of the APIs.

This API is typically implemented in a service discovery and service registry tool. When a new component is deployed to an environment, it automatically registers its endpoint in the form of an IP address or DNS name with the service registry under its unique ID (Figure 7.64.). The clients of this component reference it in their property files by ID, not by the actual endpoint. When clients of this component are deployed into the same environment, they search for their dependencies' endpoints by ID and conFigure themselves with the values retrieved from the service registry. There are multiple open source technologies on the market that provide this functionality: Zookeeper, etcd, Consul, Eureka, etc. Platforms like Kubernetes, Mesos Marathon, or Docker Swarm provide this functionality out of the box as well. Depending on the tool, the actual registration and search may be implemented in the application code (in the case of Eureka) or in deployment scripts (in the case of etcd). In either case, this functionality is implemented in the component's deployment package and executed automatically on deployment. Compared with the initial configuration, the re-configuration of application instances in run time is more complicated. It is supported by most service registry and discovery tools, but the application code itself should be implemented in such a way that it "listens" to updates in endpoint configuration and re-establishes connections when the endpoint values change.

The configuration of endpoints in the client component may also be different depending on the service component. Some components that are deployed behind a load balancer have a single endpoint, which is actually the endpoint of the load balancer. Some components may not need load balancers because the load balancing is done by client libraries on the client side. This is common with clustered system components, like databases and caches. In this case, the service registry will contain a list of endpoints for a single component.

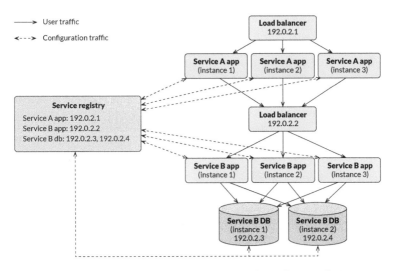

Figure 7.64. Service registry and discovery for dependency endpoints.

We mentioned above that both IP addresses and DNS names may be used as endpoints. Caution should be exercised with DNS names though, and DNS names should not become the default choice. The most important reason is that DNS names may be cached on multiple levels. Propagation of a change in cases of failover or deployment may then take time, which may lead to stability issues. Another reason is that DNS by itself can be thought of as an implementation of a service registry and discovery. Therefore, the use of DNS will lead to duplication of functionality. In some cases, however, the use of DNS is justified, but the issues above need to be analyzed for each particular case.

7.5.4. Secrets

External dependencies often require authentication on the client side. This is the best practice for most system components, like databases and caches. Therefore, in addition to the configuration of endpoints, secrets should be configured appropriately. The general workflow is the same as that in the case of service registry and discovery, but typically a different tool is used to manage secrets. A separate tool is required to handle security aspects like encryption, authentication, and authorization.

When components are deployed in the environment, in addition to registering their endpoints, they create usernames, passwords, and certificates and register them in the secret management tools. When client applications are deployed, they read the secrets and conFigure the appropriate application properties with this information.

Another use case when a secret management tool is required is the management of certificates, which is most typically used for SSL connections. Depending on the tool and the use case, certificates can be either generated by the production operations team manually and then uploaded to the secret management tool or generated by the tool itself. In either case, the resulting secrets are then pushed to the client applications and the required properties are configured on the application side.

There are various open source secret management tools. Hashicorp Vault is a good example. Platforms like Kubernetes and many cloud providers provide various secret management services out of the box as well.

Because secret management tools manage secrets, it is extremely important to conFigure them securely. Authentication and authorization is typically the most challenging aspect of such configuration, to ensure that only authorized applications and personnel have read or write access to the appropriate secrets. A description of the security details of proper secret management configuration for microservices in the cloud environment is out of scope of this book.

7.5.5. Scalability

Scalability configuration can be split into several categories, depending on the scalability direction and the type of application:

1. Horizontal scalability for stateless and embarrassingly parallel applications. In this case, the individual application instances are independent and don't have to know about each other. Typical examples of such applications are stateless business application components of services.

2. Horizontal scalability for stateful applications or applications requiring clustering. In this case, the different application instances share state or need to know about each other for various reasons. Typical examples of such applications are databases, caches, message queues, and distributed coordination and configuration systems, like Zookeeper and Consul.

3. Vertical scalability of any applications.

Only changes in horizontal scalability for embarrassingly parallel applications can safely be considered as configuration changes. Even in this case, the number of instances can only vary within certain boundaries that were pre-tested during the service changes pipeline.

Changes to the horizontal scalability configuration of clustered applications are more challenging; however, under certain conditions, these can also be considered as configuration changes. Most importantly, all possible options of configurations should be pre-tested. For example, there may be two allowed options for a Zookeeper cluster that were pre-tested: the 3-node and 5-node clusters. Changes to another set of nodes should be considered as a service-level change for the Zookeeper component and should go through more extensive testing. For example, in the case of Zookeeper, the 7-node cluster may fail performance tests because of the increased load of finding quorum, even under the same conditions as those when the 3- or 5-node clusters performed well. Even when the configuration options are pre-tested, the scaling of clustered applications should be done with caution, because in most cases, it will lead to reallocation of data across the cluster, which causes an increased load to the cluster during this process. Depending on the amount of stored data, rebalancing of the database and cache clusters may take significant time, which is oftentimes hard to predict. Unfortunately, when such a component is a system of record, there are no other options than to perform scaling in production. Such scalability changes require necessary preparations and careful analysis of the risks.

Changes in vertical scalability mean increases or decreases in the CPU, RAM, and disk space available to an application. The challenge is that such changes may lead to unpredictable behavior if not tested properly. For example, in the case of managed platforms like Java, if applications are given more system resources without appropriate tuning of other parameters like garbage collection or memory management, it may actually lead to degradation of the application performance and availability. Another challenge is that vertical scalability changes require application instances to be restarted in most cases. The only exception when a vertical scalability change can be considered as a configuration change is when various vertical scalability options were pre-tested. In our experience, this is rarely done, because the costs of pre-testing multiple options outweigh the potential benefits. When the pipeline for service changes is efficient, it is cheaper to treat vertical scalability changes as service changes and just run them through the service pipeline.

7.5.6. Traffic Routing

Depending on the type of rule, changes to the configuration of load balancers may be considered configuration or service changes. Rules that affect application behavior, like sticky sessions, routing of requests by URL patterns, or termination of SSL, are clearly service changes and need to go through the regular service changes pipeline. Changes to these rules are often hard to predict and the risk of changing them without testing is too high.

Rules related to traffic routing during upgrades or releases, however, are a configuration change. These are a change to the environment configuration, because the application itself is not aware of other versions deployed in the environment and the percentage of customers that this specific application instance should be serving.

7.5.7. Configuration Change Management

It is important to note that changes in configuration are actually considered changes only if they are performed manually by business users or production operations personnel. Because feature flags and business configuration are always manual, we discussed the fact that the business portal should have built in controls and policies for managing business configuration changes.

Other changes, like the changes to endpoints of upstream services, secrets, scalability configuration, and traffic routing configuration, can be managed in more traditional ways. These changes are considered to be configuration changes, so they are performed within pre-tested boundaries and additional testing is not required. The only remaining policy is that these changes should be authorized and auditable. The easiest way to implement such a policy is to use one of the tools that is traditionally used for production change requests. Jira would be a good candidate for such a system. For each configuration change, a ticket should be raised and approved by the operations team lead. In addition to an easy approval mechanism, Jira will provide an audit log.

However, if such configuration changes are performed automatically, they are managed differently. When automation logic included in the application deployment package is released to production, the automation has the necessary permissions to approve configuration changes implemented by this automation logic (Figure 7.65.). The automation logic inherits such permissions from the personnel that implemented this logic in the code and approved it during the service change management pipeline. Such automation logic can typically perform only configuration changes within the boundaries that were pre-tested during the service change management pipeline as well. When such automation logic performs a change, the change is effectively pre-approved by personnel with appropriate roles who initially signed off on the automation logic during the service change management pipeline. Nevertheless, even if automated configuration changes are pre-approved, an audit log is still required and is typically implemented in the automation logic directly or in the platform that executed the automation logic.

For example, if an auto-scaling policy is configured for an embarrassingly parallel application within certain boundaries, the decision to scale the application up or down is pre-approved based on certain conditions and thresholds. The auto-scaling policy is then pre-approved for any changes made within this policy by the run-time platform. Changes in the auto-scaling policy configuration should be considered as service changes and should be tested thoroughly. The original author of the auto-scaling policy and everyone who signed off on this policy during the service change management pipeline can be considered as approvers of the actual scaling decisions made by this policy. Even if the actual scaling decision is made by the platform, the chain of approvals would still lead to specific people with specific roles.

Another example can be found in service registry and secrets management. If a decision to reconFigure a client application with a new endpoint is made by the service registry and the client application supports automated configuration, such a decision is considered to be pre-approved. As in the case of auto-scaling decisions, even if the change is done by the platform, the chain of approvals will ultimately point to the specific people with specific roles who signed off on the code implementing this policy during the service changes pipeline. Changes in configuration of endpoints and secrets should be audited. Such an audit may be provided by the service registry and discovery tools, as well as implemented in the application code, which should log all changes in the application properties.

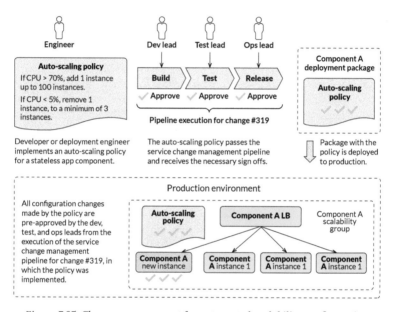

Figure 7.65. Change management for automated scalability configuration changes.

Such automation of configuration management changes allows the manual involvement of production operations personnel to be minimized and most of the decisions to be automated, while retaining an appropriate chain of approvals and an audit log.

7.5.8. Example: eCommerce Platform

All types of configuration change need to be implemented for the eCommerce platform. The tool set for managing these changes includes the feature flags management tool, service registry and discovery, secret management, and ticket management (Figure 7.66.).

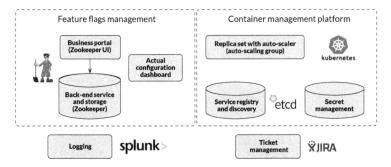

Figure 7.66. Tools for managing configuration changes for our eCommerce platform example.

Feature flags management is implemented in Zookeeper with the Zookeeper UI. Authentication and authorization is set up on the Zookeeper UI to ensure that only authorized users can make changes to the feature flags. Peer review and approvals for feature flag changes are not implemented, on the assumption that any business users will implement the necessary approval chain internally. A logging system is used for the audit log of all feature flag changes with two levels of logging:

1. The first level is implemented in the business portal, and the change is logged on the modification of a feature flag from the portal. All users of the business portal are authenticated, so both the change and the requestor are logged.

2. The second level is implemented by each application. When a property with the feature flag is changed on an application instance, the instance ID and the change is written to the application log.

The feature flags management service is a part of the production environment and is hosted in the Kubernetes cluster, together with other business and system services. Business applications are initially started with default property values for feature flags. On startup, application instances are configured with the endpoints of the Zookeeper cluster via the service registry. Application instances then connect to the cluster, read overrides of feature flags, and subscribe to notifications for future changes of feature flags in Zookeeper. As soon as a feature flag value changes in Zookeeper, all subscribed application instances are notified and change the respective property in their configuration.

In this case, Zookeeper contains only overrides of feature flag settings. If a new version of an application component is deployed to production and it contains a new feature flag, it will use the default value that is configured in the deployment package property file.

In the simplest form, the feature flags and other relevant business configuration may be structured in Zookeeper in the following way:

1. The first level contains the names of components, so that each component can subscribe only to its own feature flag changes.

2. The second level contains component-level feature flag overrides, in the form of "property_name -> property_value." The second level may also contain overrides for specific instances of a component for troubleshooting. In this case, the second level will contain the specific instance ID.

3. The third level exists only if instance-level overrides are used. In this case, it contains feature flag configurations in the form of "property_name -> property_value."

The Zookeeper UI allows business users to view and change the intended configuration of feature flags. The actual configuration that exists on the instances may be different in the case of application defects or missed notifications from Zookeeper. To see the actual configuration, each application instance should expose a REST API or web page showing the current configuration of the application instance. This API or page should include the complete configuration of the instance, including feature flags, endpoints, and other technical configurations. However, it is important not to accidentally expose secrets in plain text. Unfortunately, finding all instances of components in Kubernetes and going through them one by one may be difficult for multiple reasons. First, finding the information about all instances requires getting into the Kubernetes console. Second, it may just take too much time for highly scalable components. Last but not least, by default, these instances may not have easily accessible IP addresses and may require setting up tunnels. All of the above might make it very difficult or impossible for business users to see the current configuration. A separate dashboard automating the steps above should be implemented to show the aggregated actual information collected from the application instances. As a later enhancement, this dashboard may replace the Zookeeper UI and allow viewing and modification of feature flag configurations, so that business users have a single portal for all business configuration needs.

One of the challenges with managing feature flags in the eCommerce platform is that the platform consists of many services. When a business feature affects multiple components, it may be difficult to enable it consistently across all services and components. To solve this problem, we will assume that all composite features are managed as feature flags on the web UI component level (Figure 7.67.).

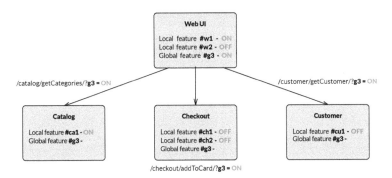

Figure 7.67. Composite features managed with feature flags for our eCommerce platform example.

The web UI component then passes overrides for the feature to the upstream services. This runtime information then overrides any configuration that may exist at the local level.

Service registry and discovery and secret management are also provided by Kubernetes out of the box. As long as Kubernetes manifests and Helm charts are used to deploy application components on Kubernetes, no additional implementation is required. In fact, manual configuration of external dependency endpoints and secrets is not allowed. The configurations of dependencies external to Kubernetes are also implemented with Kubernetes by using endpoints or services of the type ExternalName. If an application needs to be rerouted to a different endpoint, the client application property files are not changed; rather, the appropriate objects in the Kubernetes or DNS names are modified. When these modifications are required, a change request should be created and approved in JIRA.

Service discovery is implemented in Kubernetes by setting environment variables with the values of endpoint dependent services inside containers that host instances of client components. If dynamic reconfiguration of a client component is required, that client component may need to be redeployed. In the case of our eCommerce platform, DNS is not required for internal services, because blue-green upgrades would lead to the creation of the whole subgraph of downstream dependencies. However, DNS may be useful for external dependencies outside of the eCommerce platform, to avoid the need to restart or re-deploy applications or changes to external dependencies' endpoints.

Horizontal scalability configuration changes are implemented only for stateless application components, like the web UI, catalog, checkout, and customer services. The implementation is provided by Kubernetes natively in the form of an auto-scaler component attached to the corresponding replica sets. Auto-scaler configuration changes are considered as service changes, and all runtime decisions about scaling are considered to be pre-approved. Auto-scaling doesn't apply to stateful clustered applications, like databases, caches, message queues, and Zookeeper instances. If manual intervention is required for scalability decisions, the change should first be requested and approved in JIRA.

Most of the traffic routing changes for blue-green upgrades, canary releases, and A/B testing are implemented on the Akamai Global Traffic Manager level. All changes to the Akamai configuration should first be requested and approved in JIRA.

7.6. INFRASTRUCTURE CHANGES

Infrastructure changes are different from other changes, because infrastructure is managed and developed by a separate department and provided to the service development departments as a service. This means that the contract between infrastructure and applications is well-defined and most of the changes to infrastructure do not affect applications. Often, infrastructure changes are driven by special requirements related to security, cost, or just providing more features as a part of the infrastructure service.

The scope of infrastructure as we defined it before includes the compute, storage, and networking services. The compute service typically includes base images of the operating system and required system packages. We will assume cloud implementation of the base IaaS services, so internal functionality upgrades of these services are delivered by the cloud provider. Changes performed by the cloud provider are not managed by the internal infrastructure team. Participation from the infrastructure team is required for changes related to the configuration of these services, including the base OS image, networking, quotas, and access control setup.

7.6.1. Container Platforms

The exception when the compute service is managed by the separate internal team is when a container platform is implemented on top of a traditional VM-based IaaS provided by external vendors. The actual team that manages the container platform may belong to the software development department and may not be a part of the core infrastructure team. For the purposes of this chapter, we will consider changes in such a container platform as infrastructure changes. The reason is that, from the service development team's perspective, the container platform is the actual compute service, but it works with containers instead of traditional VMs.

The typical changes related to container platforms include upgrades of the platform's software, configuration of the platform to provide correct access control and quotas, and planning of the capacity on which the platform itself executes. The changes related to the configuration of the container platform follow change management policies and processes similar to those in the pipeline for the system components of a business service:

1. A separate source code repository should be created to contain the deployment scripts and baseline configuration of the platform. The deployment scripts use traditional IaaS API to deploy and upgrade the platform. Often container management platforms include baseline deployment, configuration, and upgrade scripts, so they don't have to be developed from scratch. In this case, only organization-specific configurations and overrides are put in the source code repository.

2. The build of the source code repository leads to the creation of a deployment package with the platform. The build process is similar to the build process of system components for service changes.

3. Some level of testing should be performed before the new version is deployed to production. Typically, container platforms

come with a base set of tests. For major changes, integration testing may be implemented. Integration testing would then deploy the platform to the test environment, deploy business services and components that use the platform, and run functional and non-functional test suites from business services.

4. The deployment of the platform to production is executed with a rolling upgrade. Blue-green upgrades cannot be used to maintain intact business services executing on the platform. The rolling upgrade procedure is typically supported out of the box by the container management platform, so the change comes with execution of the upgrade. To maintain the availability of hosted business services, the upgrade execution may take significant time. Each step of the upgrade typically moves all containers from a VM in the container management platform cluster, replaces the VM with the new one or just upgrades the platform software, and moves the containers back to the new VM.

For obvious reasons, all of the changes in the container management platform should be backward compatible. In the extremely rare cases of backward incompatible changes, the upgrade procedure may be extremely complex and involve data migration in system-of-record components.

Configuration changes of the container management platform, including system configuration and scalability, are managed in the same way as they are managed for application configuration changes. Even if the container management platform is a clustered application, auto-scaling may still be used if configured properly. The key is to prove that no configuration changes will affect the applications that are executed on the platform in production.

7.6.2. Base Images

Base images of the operating system and system packages may be the weakest point in the contract between the infrastructure and software development teams. Base images usually contain the version of an operating system approved by the infrastructure, as well as the proper versions of system packages that have to be a part of any container or VM instance. The reason it is the weakest point is that oftentimes there are no good ways to make updates to images and updates to the application and system components independently.

One of the ways to strengthen this point is to allow software development teams to create, maintain, and use their own images, according to the policies and requirements provided by the infrastructure team. In general, we recommend this approach. In addition to streamlining the change management process and the responsibilities between the teams, it may also help with the size of images. In our experience, one of the challenges with the images provided by a separate infrastructure team is that their size and configuration complexity are larger, just because the images have to be generic enough to support many known and unknown future uses. Images that are built by application teams are built for a specific purpose, so they are often more compact and more secure as a result of being simpler and having less attack surface. To help application teams with the creation of images, many tools to make essential application images have emerged for both container- and VM-based infrastructure.

Unfortunately, depending on the qualifications of both teams and the complexity of such requirements, this may not be always possible. In this book, we will assume that base OS images are provided by the infrastructure team to software development teams and then the software development teams have to use them to create specific images of their components. The number of base images may vary based on the organization and the requirements of the infrastructure and application teams. One reason is that the enterprise may have several approved and supported operating system versions. Another reason is to minimize the size and configuration of each base image.

There are two major types of base OS images: container images and VM images. From the change management perspective, in both cases, the base image is used as a library during the build process for deployment packages of all components (Figure 7.68.).

As in the case of application libraries, a new version of the base image should not trigger a new build of dependent components. Otherwise, in the case of widely used base images, a change in the base image may unintentionally trigger massive changes across all organizations in the enterprise. However, if this behavior is intended for some security-critical patches or some base images used by especially security-sensitive components, this configuration may be changed.

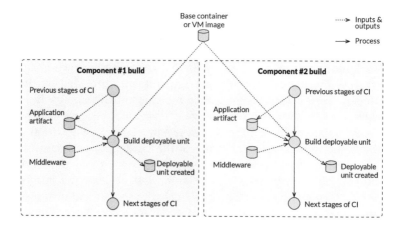

Figure 7.68. Use of a base image during the build process for deployment packages of all components

The most typical reason for changes to base images is to migrate to the new version of an operating system or the new version of a system package. To minimize the scope of downstream changes, base images should contain only the absolutely required packages. If different sets of components in the organization have different requirements for the packages of the base image, different base images should be created. That way if a component doesn't really use a system package from the base image, it won't need to be updated when a new version of a system package is released.

Whether new versions of base images trigger build and CI pipelines of downstream components or not, the infrastructure team should define SLAs to update all downstream components with the new version of the base image. The SLAs may be different for different types of change. For example, security patches will need to be applied across the board within hours. Regular updates of system packages may be allowed weeks or months to be updated. Because the base image acts as a library for the downstream components, moving a component to the new version of the image requires a change in the component's source code. Specifically, the build script of a component's deployable unit should be changed to reference the new version of the base image. Once that change is made, a new version of the component's deployable unit and deployment package will be built, tested, and prepared for release. Alternatively, the latest version of the base image can be used on every build of component. In this case, actively developed components will be migrated to new versions of the base images automatically. This behavior, however, is not always desirable, because non-critical updates of base images may interfere with critical updates of a component's functionality and block component development for a long time. The specific choices about whether take the latest versions of base images or use specific versions and whether to trigger a component's build process for every change are configurable for each component, so the trade off between speed and control can be analyzed on a case-by-case basis.

Irrespective of the exact configuration of the component's build process, the infrastructure team should have up-to-date information about what versions of base images are currently used in production by each component. To do that, the version of the base images can be embedded in the base images themselves and then this information can be obtained either from the change tracking log or from the production environment directly. In the former case, the infrastructure team can analyze what versions of components are currently deployed in production by looking into the tags and labels on source code commits and deployment package artifacts. This information can be used to find the respective versions of component build scripts and parse information about the version of the base image used to create the deployable unit. The information about the current state of versions of used base images can then be used to bring all software development teams into compliance.

The biggest challenge with VM base images is that, relative to container-based images, they contain significantly more system packages and configurations, like VM-level access control for SSH (Figure 7.69.).

In the case of container base images, the base image itself is small and doesn't contain many packages. Therefore, changes to container base images are rarer than to VM base images. For container base images, most of the changes are done on the level of the base VMs. Because the applications are isolated from the VMs by containers, VM-level changes are completely independent from application changes and can be performed in parallel by the infrastructure team with typical infrastructure configuration management tools, like Puppet.

Similar approach can be used in the case of VM images as well. Despite the larger size of VM images, the benefits of treating them as component libraries from a change management perspective outweigh the potential downsides. However, we are also going to discuss another approach to change management for VM images, which is based on runtime updates, later in this section. This approach does not represent the best practice, and we don't recommend it as a default option, but in our experience, it may still be used in some companies, so it is worth mentioning it.

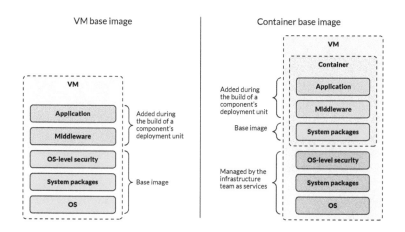

Figure 7.69. Differences between VM and container base images.

If the approach with runtime updates to VM instances is used, changes to base images are split into two separate categories:

1. Major changes that have to go through the regular process or rebuilding of the base image, rebuilding of all dependent component images, and release of all images via the service change management pipeline.

2. Minor changes that can be applied to executing VM instances in runtime. In this case, the change goes to all managed VM instances in parallel with a new base image being prepared to include these changes as well.

Major and minor changes may then be implemented in different source code repositories and different technologies to enable cleaner separation. A good analogy of such major and minor changes for base images would be the traditional view of the source code and configuration (properties) of service components (Figure 7.70.).

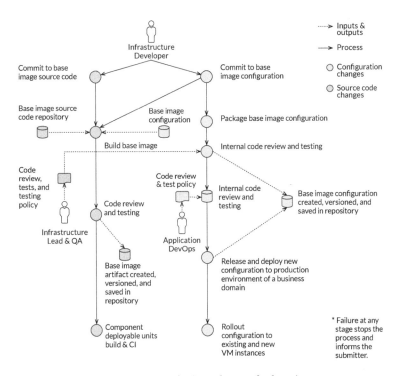

Figure 7.70. Major and minor changes for base images.

The change management process for regular base image changes that produce base image artifacts is similar to the process for service libraries. The process includes commits, code review, static code validation if needed, and testing of the new base image. The testing of the base image includes validation that the image satisfies all requirements of the component teams. It may include building representative components by using the new version of the base image, deploying those components to a test environment, and checking the base health and functionality.

Because the base image can then be used by all software development departments and service teams, those teams can implement additional validation of the base image before using it to build specific component images. This validation is implemented by using specific requirements of each department and team to avoid breaking the component pipeline if the base image is bad.

The configuration pipeline is more lightweight, but it still has the necessary controls in place. Configuration changes should pass internal infrastructure code review and testing. After that, before being released and deployed to the production environment of specific business domain departments, the changes should pass the necessary review and testing of the corresponding software development and production operations teams. Although this is a manual process that increases the time to production, it adds necessary controls that help to ensure that the changes provided by the infrastructure team are good. After the change is approved by the respective software development department operations team, it can be deployed to production. The deployment process shouldn't require VM instances to be restarted and the changes are applied to each instance in run time and at the start.

In parallel, the configuration changes also trigger the building of base images, so that the changes will already be included in the new versions of the components' deployable units the next time that they are built. To avoid double application of the configuration changes, the technologies that are used to bring the VM instances configuration to the desired state in run time should be aware of the current configuration of the VM instances. The most popular technologies, like Puppet, support this functionality out of the box. Puppet is capable of tracking the current configuration of a VM instance, so that, if the change is initially applied through the configuration pipeline but the components are then rebuilt with the new base images, the changes will not be reapplied when a VM based on the new VM image is deployed to production.

As we mentioned before, this approach allows a faster reaction to VM instance configuration changes, but it has a number of drawbacks. First, fully fledged testing of configuration changes is difficult to implement, so corners may be cut that will lead to an increased risk in applying these changes in production. Second, the additional runtime configuration of VM instances will increase VM startup times and may lead to unexpected performance degradation in run time if the configuration process applies a big bulk of changes and consumes too much of the system resources. Third, the dual change management process for essentially the same entity will lead to less control over the exact configuration in production and lead to runtime issues that are difficult to troubleshoot. Last but not least, a restart of VMs and applications may still be required to apply some of the changes.

To minimize the risks and issues, only a small set of well-understood changes should be applied with this process, and the process itself should be carefully designed. Overall, we recommend avoiding runtime updates to VMs and sticking to a more straightforward approach by treating all changes to base images as regular code changes, with the creation of base images and libraries and propagation of them to the applications with the usual change management process that is used for regular service (not configuration) changes.

The change management pipeline for base images should also include validation and testing stages. These stages can be implemented in a number of ways:

1. Base validation of the image should include tests that would deploy the image and create a VM or container instance out of it. The test may then check that the instance is healthy, that it has the required connectivity, that the security and access control is set up properly, and that the correct and up-to-date versions of packages are installed.

2. Backward compatibility and integration testing may include using a representative subset of components from various departments to build specific component deployment packages, deploy those components to test environments, execute health checks, and run smoke service-level tests.

Both approaches are better used for the base image source code changes that include the building of base image artifacts. In the case of configuration changes, a variation of the second approach can be used. In this case, before application of the configuration change to the production environment, the change can be applied to the test environment to validate that it doesn't break the services and components executing there.

7.6.3. Configuration

There are several types of infrastructure changes that are inherently difficult to test, because they are environment specific. These changes include but are not limited to networking changes and changes to quotas and access control.

The changes to networking, like creating new networks or configuring firewall rules, are difficult to test because they are environment specific. However, most of the breaking changes to networking are implemented during the initial configuration of the environment.

Late changes typically include only modification of firewall rules. If new services need to be added to the system, leading to potential changes in networking, those changes are better implemented in a backward-compatible way.

Because all networking changes are high risk, they are most often carefully reviewed, approved, and implemented in production manually, even if the actual configuration is scripted. In some cases, when subnets with firewall rules are included in the component deployment packages, the risk may be alleviated by creating new subnets during blue-green deployment.

Changes to quotas and access control follow the same principles as networking changes. Most of them are required during the initial environment setup and configuration. Ongoing changes require careful peer review because of the challenges of testing them reliably.

The environment validation testing discussed in the chapter about service testing can be used to validate the environment configuration. The challenge is that environment validation is specific to an environment, so it is difficult to pre-test certain production changes in a non-production environment.

The last potential type of change that can be considered as an infrastructure change is the actual provisioning of compute and storage resources. It is important to note that this is actually not a change managed by the infrastructure team. The infrastructure team provides the service, guarded by quotas and access control, which can be then used by the software development and operations teams. Therefore, changes related to provisioning of new resources are either:

1. Service changes, if they involve the deployment or upgrade of component and service instances.

2. Configuration changes, if they involve scaling component instances up or down.

Both types of change were reviewed in detail in the corresponding sections earlier in the book.

7.6.4. Example: eCommerce Platform

Our eCommerce platform is deployed with Docker containers on the Kubernetes container management platform, which helps to keep base images small and minimizes the rate of changes to the base images. The original VM-based infrastructure service is hidden by Kubernetes, and Kubernetes is the IaaS from the software development team's perspective.

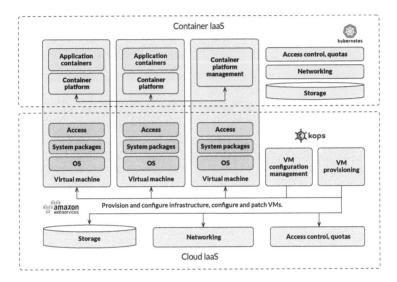

Figure 7.71. Infrastructure for our eCommerce platform example.

Because the VMs are hidden from the software development departments, updates and changes to VMs are managed separately by the infrastructure team. Kops is used for provisioning, maintenance, and upgrades of the Kubernetes cluster. In fact, AWS is going to release managed Kubernetes clusters as a service (EKS); when this happens, it might make sense to transition to the managed offering.

Kubernetes deployment and configuration scripts are stored in specific source code repositories. The change management pipeline for all repositories includes the typical steps of code review and testing. Most of the changes can be applied without special approvals from the software development team, with the exception of major high-risk changes, like major upgrades of the Kubernetes cluster. In general, upgrades to infrastructure should be aligned with periods of low usage of the platform by the customers. For example, in the case of our eCommerce platform, these upgrades can be done during the night on weekdays.

To verify configuration changes to the infrastructure, VMs, and container platform, the changes are first rolled out to test environments of the eCommerce platform and validated with the environment validation test suite developed by the eCommerce platform team. Environment-specific configurations are applied to the corresponding environments without prior testing and only after peer review.

A base container image with the approved version of the operating system and up-to-date versions of system packages is provided by the infrastructure team to the software development department. The base image includes only the minimum required set of packages that are common and used by all components. Changes to the base container image are rare and don't automatically trigger the build and CI pipelines of the dependent components. When a new base image is ready, the component's development team should change the deployable unit build scripts to use the new version of the image. To ensure that the base images are good, the eCommerce platform team has a set of base image validation tests that is executed every time that the infrastructure team provides a new version of the base image. The image is labeled to be approved for usage by the eCommerce team only after it passes the validation.

Information about the version of the base image is included in the image itself and in specific components' images, so the infrastructure team can monitor the state of compliance of deployed components with SLAs to migrate to new versions of the base image.

8

ENVIRONMENT MANAGEMENT

In the traditional development of monolithic applications with manual provisioning and deployment, the term environment was used to describe both the instance of an application and the actual environment that the application instance was executing in. For a microservices architecture with full automation of provisioning and deployment and a dynamic infrastructure, this term should be clarified.

As we mentioned in the chapter about service deployment, the environment is a combination of services with a well-defined contract that application instances require for execution. Major services that the environment provides include infrastructure, external dependencies, data, and application configurations. All of the services should be exposed over a well-defined contract, so that the application instance can be built once and deployed into any environment without modifications or major changes in functionality. A strong contract between the environment and application, as well as repeatability of application functionality across environments, ensures that successful testing of an application in one environment guarantees that the same application will work well in different environments.

8.1. ENVIRONMENT BOUNDARIES

One of the challenges that traditional development teams may face with migration to the microservices architecture is the choice of the right boundaries between the environment and service instances. Because the environment includes external dependencies as a part of its contract, it may be tempting to create separate

dedicated environments with dedicated instances of upstream services for each service. This may become a common approach when
service-level testing is performed with real instances of upstream
dependencies instead of stubs. Initially, it may seem like a good idea
when it is taken into account that deployment and provisioning are
automated and such environments can be created on demand for
each new service instance. However, two major downsides of this
approach are the extremely high infrastructure costs and the high
maintenance and support costs, because in spite of automation,
something may still go wrong. To avoid the downsides, environment boundaries should be carefully planned from the perspective
of external dependencies.

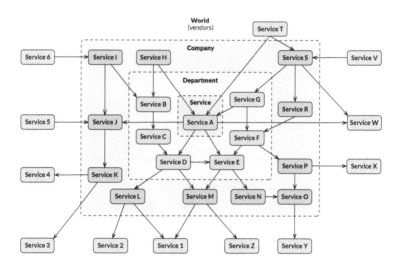

Figure 8.1. Planning environment boundaries

For example, in the case shown in Figure 8.1., if a new environment with new dedicated instances of services D, E, L, M, N, and
O needs to be created for the testing of service A, the infrastructure costs for the creation of such environments and the support
costs for maintaining these environments will be extremely high.
Essentially, if a company has n services dedicated instances of dependencies need to be created for the testing of each service, then
the total number of service instances will be $O(n2)$, which is very
expensive.

There are three ways to minimize the cost:

1. Use mocks and stubs for external dependencies. Unfortunately, the creation and maintenance of mocks and stubs may be high as well. In some cases, integration testing will still be required, so the mocks and stubs will also lead to longer testing pipelines.

2. Use dynamic instances, provision new infrastructure only when needed for testing or experimentation, and destroy when no longer needed.

3. Use shared instances of external dependencies.

The third approach is already used for all truly external dependencies, that is, for the services provided to the company by external vendors. In some cases, vendors allow the creation of test versions of their services and then the test versions are reused across all environments. The same approach should be taken, at least, when managing dependencies between departments. In Figure 8.1., this means that when doing service-level or integration testing of any service within the department, shared instances of services L, M, N, and P should be used. In turn, the department should provide shared test instances of services B, G, and F to other departments. These test instances of services should be maintained and upgraded after new versions of the services are released to production.

The same rules should be applied for service-level testing within the department. That is, the development teams of respective services, the release engineering team of the department, or the site reliability team of the department need to manage shared test instances of services that can be used for service-level testing of other services. Dedicated instances of services will still need to be created for integration testing within the department, but the use of shared services whenever possible allows the total number of service instances used for testing in the company to be minimized from $O(n^2)$ to $O(n)$.

The use of shared instances also means that new environments don't have to be created every time that a new version of a service is tested. Shared instances of external dependencies become a configuration of the environment, and then only new instances of the services under test are created and upgraded within the same environment.

8.2. ENVIRONMENT TYPES

Different environments are defined by the specific implementations of the contract that they provide to the services deployed in them. The creation and maintenance of different implementations of infrastructure, dependencies, data, and application configurations leads to higher cost, but this cost is usually justified because it helps to save on infrastructure costs by deploying service instances in smaller configurations or to increase the efficiency of development and change management by providing higher flexibility and relaxing security and access control.

The typical environments used in development and change management include the development sandbox, service-level testing, optional integration testing, performance and stability testing, and production. We do not include a separate staging environment, because we assume that blue-green deployment can be used for major upgrades and business acceptance testing can be done in production before the production traffic is switched to the new version. In this case, staging is not a separate environment but an instance of a new version of a service in production. Table 8.1. describes the different non-production environments.

Table 8.1. Non-production environments.

Environment	Description	Ownership and Access
Development sandbox	Includes on demand sandbox instances of services for experimentation and testing by the development and deployment teams. * It is a separate isolated sandbox from the infrastructure perspective, typically implemented as a separate cloud project. The cloud project may be shared across different service development teams belonging to the same department. * Service virtualization or shared instances of external dependencies and test data from testing environments or manually created data are used. * The application and environment configuration is highly dynamic to allow experimentation. Instances are typically small, but they may be large to experiment with vertical and horizontal scalability.	Owned by service-level development teams. The teams have wide permissions for the infrastructure and full control over the application configuration.

Environment	Description	Ownership and Access
Service-level testing	Includes functional and profiling instances of services, which are deployed during the service testing pipeline. * Infrastructure can be shared with other testing environments. * Service virtualization or shared instances of external dependencies are used. * Small service instances, just large enough to support the execution of functional tests. Medium-sized instances are required for profiling testing.	Owned by the respective service-level QA teams. The QA teams have permissions to change the application and environment configurations.
Integration testing	Includes instances for global-version integration testing. If all changes are independent, this may not be required. * Infrastructure can be shared with other testing environments. * Real dependencies of services are included in the global version (developed inside the same department). Shared test instances of external out-of-department dependencies are used. * Small service instances, just large enough to support the execution of functional tests.	Owned by the QA team that provides integration tests. The QA team has permissions to change the application and environment configurations
Performance and stability testing	Includes instances for performance, stability, and other types of non-functional testing. * Infrastructure can be shared with other testing environments. * Real dependencies of services are included in the global version (developed inside the same department). Shared test instances of external out-of-department dependencies are used. A different set of shared external instances may be required to support high loads. * Large, production-like service instances.	Owned by the performance QA team. The QA team has permissions to change the application and environment configurations.
Production	Production and staging instances of all services within the department. * It is an isolated production infrastructure with limited access. * Real production endpoints of all external dependencies are used. * A large scale is required to service production traffic.	Owned by the site reliability engineering team.

The environments described above may be split further into more fine-grained environments. The guiding principle is that, if different environments require different access controls or environment configurations including external dependencies and infrastructure, then it may make sense to split them into different environments. Typically, the number of environments should roughly correspond to the stages of the change management process and to the different teams who are responsible for specific stages of testing.

Each environment may host any number of actual instances of the system under test. However, the boundaries of the system under test are different for each environment. That is why, for example, the service-level testing environment and integration testing environment are split. The service-level testing environment will allow the creation of instances of specific services, and all upstream dependencies of that service will belong to the environment configuration. The integration testing environment will allow the creation of sets of instances of services that belong to the same business domain or are developed and tested by teams within the same software development department. Only dependencies that are external to the set of services as a whole will belong to the environment configuration; every internal service will be created when an instance of the system under test is created.

8.3. ACCESS CONTROL

The ownership of the environment means that the owner team does not only have access to change the environment configuration and deploy and upgrade instances of services. The owner team has a responsibility for supporting the environment and instances of the services executed there. In the case of misconfiguration, defects, or instability, the owner team should troubleshoot the issues. The owner team may involve other teams, like development, deployment automation, or infrastructure, to help with the troubleshooting. However, the owner team is ultimately responsible for the stability of the service instances and the environment as a whole.

In addition to ownership of the environment, it is important to define ownership of the instances deployed into each non-production environment. The typical policy is that whoever created an instance has access to the configuration and destruction of that instance. If an instance is created automatically during the change management process, the release engineering team owns that instance. Unfortunately, the configuration of this policy is challenging unless a separate self-service interface is implemented to manage environments and service instances.

A simpler policy that is typically implemented gives the release engineering team full access to all instances in all non-production environments, except the development sandboxes. A number of engineers from the service development, deployment automation, and testing teams, who have passed special training, are also given full access to all instances in the non-production environments to allow troubleshooting. Training is required to ensure that the engineers don't make permanent manual modifications to instances and that they instead destroy instances after troubleshooting and modifications, introduce changes via the regular change management process, and allow the CICD pipeline to recreate instances appropriately. Other team members only have read access to avoid them tampering with instances that are used to sign off changes and to guarantee that the test instances are indistinguishable from future production instances of that version. For sandbox environments, access control is less strict; all development team members can have full access to instances hosted in sandbox environments.

8.4. ENVIRONMENT CONFIGURATION

Ownership of the environment is effectively defined by the team that controls the environment configuration, including the infrastructure requirements, specific endpoints and configuration of external dependencies, data, and environment-specific application configurations, like scalability and feature flags.

The production environment configuration and changes to this configuration were discussed earlier in the book. Obviously, the production environment is highly protected, and each change should go through the required review and approval process and then be saved in the audit log.

Environment configuration for non-production environments is important as well. The configuration of each environment should be stored and versioned in a separate source code repository. Depending on the importance of the environment, minimal change management should be implemented for non-production environments. The process can be as lightweight as a basic peer review and approval.

The only exception from the rules above is a dynamic environment configuration that may be automatically changed during runtime. For example, the scalability configuration doesn't need to be versioned when auto-scaling is enabled, because it will be changed by the platform in response to changes in load. The endpoint configurations for internal services and components also don't need to be versioned, because they will be automatically updated by the service registry and secret management tools. However, the endpoints of external dependencies need to be stored as a part of the environment configuration and versioned.

8.5. SELF-SERVICE

The self-service tool is an important feature to enable efficient management of environments and service instances inside environments. In many cases, the lightweight implementation of self-service can be provided through source code repositories, in which the environment configuration is stored, deployment scripts in the deployment packages of individual components, and the interface that the infrastructure-as-a-service provides. Most public and private cloud vendors, as well as container management platforms, provide both console and web user interfaces to monitor and manage instances of deployed applications. In all cases, however, these interfaces are quite technical and focus on managing the underlying resources, like load balancers, VMs, containers, scalability groups, storage, and networks. If a high-level service-oriented view is required, additional proprietary tools may be considered or home-grown tools may be implemented from scratch.

Self-service may also help with enforcing access control and ensuring that only a subset of changes are allowed, because working instances of services can be hidden from users to avoid random tampering.

8.6. DYNAMIC INSTANCES

One of the approaches to save infrastructure costs is to implement dynamic provisioning and destruction of service instances. New instances should be provisioned only when needed for experimentation or during the testing pipeline, and they should be destroyed when the testing finishes and they are no longer needed. We assume that the provisioning and deployment of service components is fully automated, so the implementation of on-demand service instances is not difficult and can be achieved almost by default. Although this approach brings significant benefits for the development-sandbox type of instances, the savings for testing instances may be negligible or less than initially projected.

To achieve savings in the case of both sandbox instances and instances created during the change management pipeline, the instances should be automatically destroyed after a period of inactivity. Given a self-service tool to create instances on demand, team members may readily use this functionality to create new instances but forget to destroy them at the right time. Therefore, an automatic destruction policy should be implemented. When a team member creates a new instance, they should specify how long the instance is required for. When the requested time expires, the instance should be automatically deleted if the team member hasn't explicitly requested an extension. Extensions may also be limited to a certain period of time to avoid too long leases and to put a cap on the maximum lifetime of an instance. The achievement of savings for instances created during the change management pipeline is more difficult and several factors should be taken into account.

First, if the testing stage is unsuccessful, the running instance of a service may be needed for troubleshooting. If dynamic instances are implemented for testing, this case should be taken into account and the pipeline should not destroy the instance automatically after the failed testing round. This may lead to an accumulation of failed instances and increased infrastructure costs. One of the approaches to overcome this problem is to destroy the actual instance, but to save all test execution logs, application logs, and infrastructure logs. If logging is implemented carefully, the information in the logs may be sufficient for troubleshooting.

Second, if testing is taking more time than it takes to implement and commit a new change to the source code repository, the testing pipeline may be designed to execute in a loop. The new round of testing may start as soon as the previous round finishes, which will lead to deployment and test execution without delay, with service instances required all the time. If the development teams are split between geographical locations, it may be the case that testing will run around the clock during working days. The only exception may be weekends, when the development teams are not working. If dynamic instances are required all of the time, preordering a number of reserved instances from the cloud providers may provide higher savings than optimizing the creation and destruction of service instances during inactivity.

Last but not least, in the projection of savings and implementation of dynamic instances, the time to provision resources for the instance, populate data, and warm up caches should be taken into account. The time to destroy the instance should also be calculated. If the total time spent on provisioning and deprovisioning is high relative to the useful time of testing and to the time when an instance is not needed, it may make more sense to keep instances at least partially alive. This may not only save total cost but may also decrease the total time of testing and total length of the change management process. Keeping a number of service instances alive makes the most sense for databases and caches, especially if either the volumes of data are high or the process to populate data is complex and takes significant time.

8.7. EXAMPLE: ECOMMERCE PLATFORM

As we described in the chapter about service changes, the testing pipeline of our eCommerce platform includes service-level testing and global-version integration testing. We also assume that, even if the company has other software development departments , the eCommerce platform and eCommerce platform services may have external dependencies. Other departments provide endpoints for their dependencies that can be used for functional and performance testing. Separate endpoints are used for the production environment. External dependencies are considered in the same way as services provided by vendors outside of the company, so we will not focus on describing how instances of those services are managed by other departments. With all this taken into account, Table 8.2. Shows the environments that are created for the eCommerce platform.

We will not focus on a description of the production environment, because the eCommerce platform production deployment was covered in detail earlier in the book. Instead, let's review the environment configuration and instances for the development sandboxes (Figure 8.2.).

Table 8.2. Environments in our eCommerce platform example.

Environment	Description	Ownership and Access
Development sandbox	One environment for the entire eCommerce department. Allows the creation of on-demand sandbox instances of all components and services including the web UI, catalog components, checkout components, customer components, and system components. Uses stable versions of services in the integration testing environment as external dependencies when needed. Developers may conFigure services to work with real instances of dependencies inside the sandbox environment.	Owned by the development teams of the respective services. The teams have full access to the environment and application configurations and to deployment, which is protected with quotas to control infrastructure costs.
Service-level testing	Separate environments for each service: * Web UI, * Catalog service (functional and profiling), * Checkout service (functional and profiling), * Customer service, * Platform components. Environment for each service creates full service instances with all service components. Environments are configured with proper shared instances of services and components in the integration environment. Some environments, like checkout. are configured with stubs instead, because checkout service-level testing is performed with stubs. For example, the catalog service instance includes * Catalog business application component, * Import business application component, * Catalog DB. The Kafka message queue is considered a shared component in this case. The catalog environment points to a shared Kafka queue in a stable instance in the integration environment.	Owned by the service-level QA teams. The teams have permissions to manage the application and environment configurations.
Integration testing	Allows the creation of instances for functional integration testing of the global version and for security testing. Security testing is performed in this environment because large instances are not required for this type of testing. The environment is configured with special test Akamai endpoints to allow end-to-end testing through Akamai. Also contains stable instances of services that are used for sandboxes or service-level testing.	Owned by the release engineering team, which collects requirements from the web UI and security testing teams. The team have permissions to manage the application and environment configurations.

Environment	Description	Ownership and Access
Performance and stability testing	Allows the creation of instances for performance, stability, and security testing. Uses specific endpoints of external dependencies that support performance testing. The scale of each component is production-like to ensure that the components will support high loads. The environment is configured with special test Akamai endpoints to allow end-to-end testing through Akamai.	Owned by the performance QA team. The team has permissions to manage the application and environment configurations.

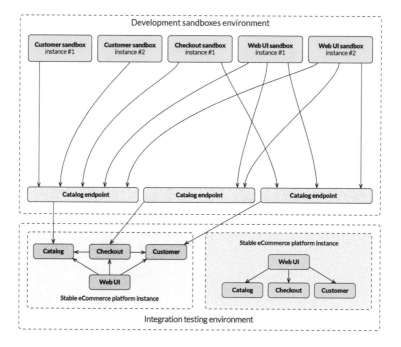

Figure 8.2. Environment configuration and instances for the development
sandboxes in our eCommerce platform example.

By default, the environment is configured in such a way that individual development sandbox instances of services are configured to work with stable versions of upstream services. Instances with upstream services are deployed to the integration environment.

However, the development teams have a great degree of flexibility to conFigure sandbox instances of services. If there is a need, development teams can conFigure their services to work with other sandbox versions of services.

The static configuration of each environment is stored and versioned in Git repositories with a basic code review. Self-service is provided via a number of basic Jenkins jobs that execute deployment scripts against the correct environments. For the development sandbox environment, the development and deployment automation teams also have access to the Kubernetes console for finer-grained management of the environment and service instances. Higher-level environments used in the change management process are protected from modification by teams to avoid untraceable changes. Only certain roles in the organization, like release engineering and select members of the service development, testing, and deployment automation teams, who have passed special training, have access to the change environment directly via the Kubernetes console.

9

MICROSERVICES PLATFORM

Earlier in the book, we discussed the fact that traditional platform-as-a-service is challenging to implement at scale in enterprise environments. One of the key challenges is the poorly defined interface between the application and the platform, which leads to leaky abstractions and the lack of reusability of the platform services as a result. However, there is a set of well-defined services that is required to implement a microservices architecture with a good change management process. We will call this set of services collectively a microservices platform.

9.1. CORE PLATFORM

Let's start with the core platform and add the services for the change management pipeline later. The core platform provides a set of application-independent features that every application needs. Each feature, combination of features, or platform as a whole can, in turn, be implemented with a proprietary open source tool and provided as a service to the application development teams (Figure 9.1.).

All of the features shown in Figure 9.1. can be provided as a service with a well-defined contract. The configuration of the respective service may still be application- or domain-specific, but even this configuration has well-defined boundaries. In comparison with traditional PaaS offerings that get into application technology domains, like middleware, databases, and caches, the platform above leaves the implementation of these aspects to the application development teams. This helps to reinforce the contract and boundaries between the application and the microservices platform, and it prevents significant changes in the platform when the application has specific requirements for database or cache configurations. An additional piece of evidence that each of the services has a well-defined contract is that many of them are already successfully provided as a service by various vendors and are successfully used at scale. In a sense, all of the services in Figure 9.1. can be treated as an extended infrastructure-as-a-service.

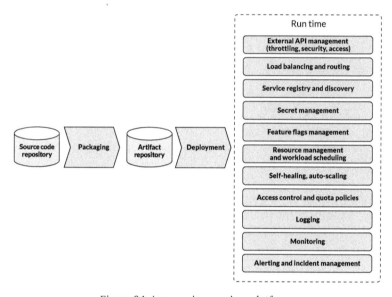

Figure 9.1. A core microservices platform.

The microservices platform can be implemented in a number of ways. The major determining factor for the technology stack is the packaging mechanism, that is, the format of the deployable units. We will focus on open source stacks for the two major options of container packaging and VM packaging. For container packaging, we will take the Docker and Kubernetes stack as an example, and for VM packaging, we will discuss the Hashicorp suite. We will not consider other packaging methods, like Java war or ear files, because they are middleware, application specific, and often too constraining for modern architectures implemented in different programming languages, technologies, and middleware.

Some of the technology choices mentioned in Table 9.1. are the same for both stacks because, at the time of writing, neither the Kubernetes or Hashicorp stacks provide full functionality out of the box.

Table 9.1. Selected technologies for microservices platform implementation.

Feature	Container-based stack	VM-based stack
Source code repository	Git	Git
Packaging	Docker	Packer
Artifact repository	Docker Registry	Native cloud storage (AWS EC2, Google GCE)
Deployment	Kubernetes + Helm	Terraform
External API management	Apigee, Mashery, custom	Apigee, Mashery, custom
Load balancing and routing	Kubernetes + cloud native, HAProxy, nginx	Cloud native or custom (AWS ELB, Google L7 or L4 LBs) (HAProxy, nginx)
Service registry and discovery	Kubernetes	Consul
Secret management	Kubernetes	Vault
Feature flags management	Zookeeper, Spring Cloud Config, Custom	Zookeeper, Spring Cloud Config, Custom
Resource management and workload scheduling	Kubernetes	Cloud-native + Nomad
Auto-scaling, self-healing	Kubernetes	Cloud-native (Scalability groups)
Access control and quota policies	Kubernetes	Cloud-native
Logging	Kubernetes + ELK, Splunk, Sumologic, or any custom	Cloud-native + ELK, Splunk, Sumologic
Monitoring	Kubernetes + Datadog or any custom	Cloud-native + Datadog or any custom
Alerting and incident management	PagerDuty, ServiceNow	PagerDuty, ServiceNow

In spite of the fact that one stack works with containers and the other with VMs, there are more similarities between them than differences. For example, the building and deployment of VM images with Hashicorp stack and Google Cloud is surprisingly similar to the building and deployment of container images with Docker and Kubernetes. This may be because Google Cloud is internally implemented in the Borg-like infrastructure that inspired Kubernetes, or it may be for some other reason. In any case, the Hashicorp stack, similar to Kubernetes, is cloud agnostic, so migration between cloud providers is manageable. Most popular cloud providers, like AWS, Google Cloud, and Microsoft Azure, are capable of providing the necessary foundational functionality to implement a microservices platform on top.

Similar microservices platforms may be implemented with other technologies. For example, a container-based stack may also be implemented Mesos DC/OS, Docker Swarm, or container-management platforms provided by cloud vendors. In some form, a VM-based stack may also be implemented with Ansible or Chef.

Other features that were not included in the template above are upgrades and canary releases. We assume that these features are implemented by a combination of deployment and load balancing & routing services. At the moment, we find it difficult to define these features as separate services, rather than business workflows on top of existing services. Canary releases may also require some form of A/B testing, which may be outside of the scope of the microservices platform.

We also don't include features related to internal API management, including throttling, circuit breakers, and API security. These features are important for microservices architecture, but in most cases, they are embedded in the services themselves with various libraries and are often technology specific. For example, Netflix Hystrix and Spring Cloud are specific to Java stacks.

9.2. CHANGE MANAGEMENT PLATFORM

The change management platform focuses on the implementation of the process itself, as well as providing environment management and test execution platforms for both functional and non-functional tests (Figure 9.2.).

Unlike the core microservices platform, the change management stack doesn't diverge into different implementations based on any deciding factors, like the packaging mechanism. There are tool suites, like the Atlassian stack, that are pre-integrated and cover many aspects of change management, but there are no tools that provide implementation of all required features out of the box. Table 9.2. provides examples of implementation options for the change management platform.

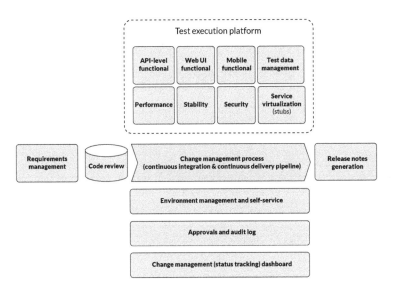

Figure 9.2. A change management platform.

Table 9.2. Selected technologies for change management platform
implementation

Feature	Implementation options
Requirements management	Atlassian JIRA, Thoughtworks Mingle
Code review	Gerrit, GitHub, GitLab (Bitbucket)
CICD pipeline	Jenkins, Atlassian Bamboo, Thoughtworks Go!
API-level functional testing	Fitnesse
Web UI functional testing	Selenium
Mobile functional testing	Selenium, Appium, device emulators
Performance testing	Jagger, JMeter, Gatling
Stability testing	Netflix OSS Simian Army + custom
Test data management	Custom
Service virtualization	Wiremock, custom
Release notes generation	Custom, GitHub
Environment management	Custom
Approvals and audit log	ELK stack, custom
Change management dashboard	Atlassian stack, custom

For the test execution platform, the focus is not to provide guidelines on the technology choices to implement tests. The focus is on the technologies to execute tests. The analogy with the core platform is that the tests are source codes that are packaged and deployed for execution into the runtime platform. The platform itself provides an environment and a service to accept a package with tests and execute it. The same considerations apply to the other features of the platform.

9.3. PLACE IN THE ENTERPRISE

Irrespective of whether the microservices platform is defined as a separate service, all of its features are required for implementation of business applications with microservices architecture in the cloud with efficient change management. The implementation may be done by the respective service development teams, dedicated teams within each software development department, or on the enterprise level. To avoid duplication of effort and lack of standards, but to give enough flexibility, we recommend that this platform is implemented either on the software development department level or as an extension of the infrastructure department. In some cases, a separate enterprise-wide department may be created to implement and provide such a platform as a service to other departments.

As in the case of the base infrastructure-as-a-service, a build-versus-buy decision must be made. Unfortunately, there is no single vendor at the moment that would provide all features of the platform in an integrated fashion. Cloud vendors, like Amazon, Google, or Microsoft, come closest to providing such functionality as a service. Unfortunately, their interfaces are still too proprietary and non-standardized, so migration between the platforms provided by different cloud vendors will be challenging. If a long-term lock-in to a single cloud vendor is compliant with the corporate strategy, then a platform provided by a major cloud vendor may be a good option.

Irrespective of how much the company relies on external vendors, as in the case with traditional IaaS, an internal team should be responsible for building, integrating, or adopting existing solutions to the company's needs. In the build case, the team may be responsible for implementation of a platform based on open source technologies. In the buy case, the team may be responsible for integrating different solutions and configuring access control policies, usage quotas, and configuration management policies.

9.4. EXAMPLE: ECOMMERCE PLATFORM

The microservices platform for our eCommerce example is implemented on the basis of open source technologies (Figure 9.3.). The core platform is based on Docker container packaging and utilizes as many Kubernetes and cloud-native features as possible.

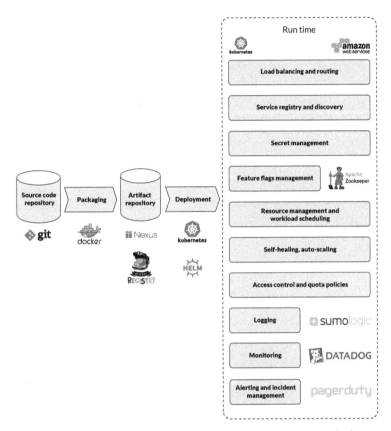

Figure 9.3. The core microservices platform for our eCommerce platform example.

The services in the eCommerce platform do not expose APIs externally, so API management is not used. Load balancing is provided by Kubernetes internally, in combination with Amazon elastic load balancers. Service registry and discovery, as well as access control, quotas, resource management, and container lifecycle management, are provided by Kubernetes out of the box. Feature flags management is implemented with Zookeeper and Zookeeper UI. Logging and monitoring is implemented by integrating with Sumologic and DataDog. Alerting and incident management is done by integrating PagerDuty with monitoring and logging tools.

The change management platform is implemented on the basis of an open source stack, except that JIRA is used for requirements management (Figure 9.4.).

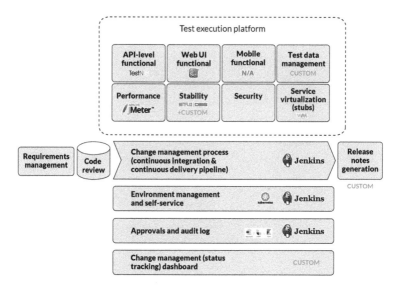

Figure 9.4. The change management platform for our eCommerce platform example.

The code review is implemented in Gerrit, and the end-to-end CICD pipeline is implemented in Jenkins. Jenkins is also used to provision development-sandbox instances on a self-service basis. The approvals and audit log is implemented with the ELK stack of Logstash, Kibana, and ElasticSearch. The test execution platform includes a Selenium cluster and TestNG for functional testing, JMeter for performance testing, Wiremock for service stubs and mocks, and a Netflix stack with custom modifications for stability testing.

Apart from open source tools, several pieces of the platform are implemented with custom code, including the test data management platform, release notes generation scripts, and change status tracking and reporting dashboard.

———————

10

METRICS AND KPIS

The change management process that was described in the previous chapters can be fully automated but still remain inefficient. To become efficient, each individual step of the process and the process as a whole should be optimized. Optimization starts with defining and measuring the key metrics and selecting the right KPIs. The change management process is implemented on the department level, so some of the metrics should be implemented for the end-to-end process and become KPIs for the head of department, the lead, and all teams within the department. Some metrics should be implemented for specific services and become KPIs for the respective service teams.

All metrics and KPIs are split into a number of categories:

1. Performance: This category includes metrics related to latency and throughput of processing changes. The latency, or speed, determines how fast a single change can be processed by the change management process. The throughput defines how many changes can go through the process end-to-end over a period of time. These metrics are similar to the regular performance metrics of any application.

2. Quality: If the change management process has a high performance but allows too many defective changes to be deployed in production, the process needs improvement. Quality metrics define how efficient each step of the pipeline is in finding defects and stopping bad changes.

3. Cost: The change management process may have high performance and catch all defects before production, but it may be prohibitively expensive. The cost category of metrics is usually

split into infrastructure costs and team effort costs to design, implement, and maintain the process.

A combination of metrics from all categories provides an aggregated metric that measures the overall efficiency of the process. In addition to the three major categories above, efficiency may also have components that are harder to measure, such as team motivation and satisfaction, time wasted on context switching, etc.

The visibility, transparency, and control of the process are not defined as metrics for several reasons. First, these are hard requirements for the process, rather than metrics and KPIs that can be optimized. The process and tooling should provide sufficient visibility, transparency, and control; otherwise, the process is not good. Second, it is hard to measure visibility and transparency, because they are qualitative and subjective, rather than quantitative.

10.1. PERFORMANCE

Performance metrics are implemented on the basis of the timestamps for the start and finish of each step of the process. The raw data about the start and finish times can be collected from the audit log, from the tool that implements the pipeline, or in a separate database for ease of calculations at a later date. In addition to timestamps, information about whether the step succeeded or failed and information about previous steps in the pipeline should be collected as well. Optionally, information about the original submitter and approver of the change in the source code repository may be collected as well. The good news is that if change status tracking is implemented properly, the required raw data will already be available in the system. With the most basic approach, the information available in change management process implementation tools, like Jenkins, will be sufficient to calculate basic performance metrics.

The following performance metrics should be calculated:

1. Time from the beginning of the step to the end of the step. This is the latency metric.

2. Number of individual changes that went through the step during a period of time. This is the throughput metric. The period of time should not be too small. Periods of 24 hours or 7 days are good for the calculation of throughput.

3. Success ratio indicating how often changes were approved for the step and promoted to the next step.

Performance metrics should be collected for each individual service pipeline and calculated for the following steps:

1. End-to-end process.

2. Each step of the process including
   a. build and packaging,
   b. deployment of a new service instance to the service-level testing environment, including environment validation,
   c. each stage of service-level testing, including smoke testing, full functional testing, and profiling testing,
   d. deployment of a global-version service to the integration testing environment, including environment validation,
   e. each stage of integration testing, including smoke testing, full integration testing, performance testing, stability testing, and security testing,
   f. final preparation of release artifacts, including the release notes collection,
   g. deployment to production.

3. In some cases, groups of steps, such as:
   a. end-to-end service level testing,
   b. end-to-end global version testing.

The collected metrics should then be grouped by department and service team. Optionally, they may be aggregated by the original submitter of the change, although these data need to be treated with care to avoid blame. The usual statistics should be calculated for each group, including the average, median, and standard deviation.

The metrics for the end-to-end process then become KPIs for the whole department and release engineering team. The end-to-end metrics calculated for each service become KPIs for the service development teams. The metrics for specific steps are used to tune the process but do not necessarily become KPIs for individual roles within teams, to make sure that the whole team has a common goal.

Metrics related to the probability of success will be discussed separately in the Quality section. They don't have to be used as performance KPIs for teams, but they may be helpful to design the process in a way that would provide the fastest possible feedback to the development teams and optimize the total throughput of the process.

10.2. QUALITY

The top-level quality metric is the number of defects found in production. Unfortunately, it is difficult to implement this metric with the change management toolset. Rather, it should be measured by the number of defects found by end customers of the platform and by the number of hotfix requests. The number of hotfix requests can be measured based on the number of the tickets for the type "defect" in the requirements management tool and by the number of commits made in hotfix branches. The former is relatively easy to calculate; however, the defect ticket should have a special configuration so that the original requestor can specify whether the defect was found in production or during development and testing.

In addition to the ultimate quality metric, intermediate quality metrics can be calculated by using the success rate metrics discussed in the Performance section above. The success rate metrics can be used as quality metrics in the following way: the combined failure rate of all future steps becomes a metric for either original submitters of the change or approvers of the previous step. For example:

1. The combined failure rate of all testing steps of the pipeline is a quality metric for the development and deployment automation team.

2. The combined failure rate of the integration testing steps of the pipeline is a quality metric for the service-level testing team.

The metrics can be more fine grained: for example, the failure rate of the full service-level testing becomes a quality metric for the smoke testing stage.

Essentially, just as the ultimate quality metric is the number of defects found in production, an intermediate quality metric is the number of defects found by the next stages of the pipeline.

Similar to performance metrics, quality metrics should be aggregated by department, service team, and (optionally) the submitter. The usual statistics should be calculated for each group, including the average, median, and standard deviation.

The end-to-end ultimate quality metric aggregated on a department level becomes the KPI for the entire department. The end-to-end ultimate quality metric aggregated by services becomes the KPI for the entire services team. The intermediate metrics for specific steps should be used to optimize the process, as we discussed earlier in the book in the section dedicated to pipeline optimization.

10.3. COST AND EFFICIENCY

Cost metrics come in two different flavors: the cost of work or human resources and the cost of infrastructure resources.

The cost of work is difficult to calculate automatically. As an option, it may be calculated as the ratio of supporting roles to value-generating roles. For example, the ratio between test engineering and developers or the ratio between release engineers and developers. This, however, may be a controversial metric by itself, because if it becomes a KPI for certain teams, management will have motivation to just formally rename the roles, although the amount of supporting work may still be high. This, in turn, may lead to motivation issues and hidden costs. Typically, the actual cost of work may be indirectly inferred from other metrics. For example, if the time to get a single change though the pipeline is long and the end-to-end success ratio of the pipeline is small, especially when the majority of failures are discovered at the end of the pipeline, it may often lead to significant inefficiency and wasted effort as a result of context switching. The time that it takes to develop the smallest changes may help to calculate the cost and efficiency of processing a change. For example, if every change requires rework of the majority of tests, creation of significant test data, and rework in stubs and mocks, it may mean that the test strategy and design is inefficient and requires adjustment.

The infrastructure costs are easier to calculate. Significant infrastructure costs come from the test environments, and calculation of the total cost of the change management infrastructure is not difficult. With proper tagging of cloud resources, the total cost per department and per service can be calculated by the cloud itself and reports can be generated on a periodic basis. The coarse-grained infrastructure cost metrics should include:

1. The cost of the change management platform, including source code repositories, artifact repositories, CICD pipeline

management tools, test execution clusters, and change tracking dashboards.

2. The cost of service instances in non-production environments, which should be calculated by environment, service, and component. This is easy to implement with proper tagging of cloud resources when a new instance is provisioned during the change management pipeline execution.

In addition to the metrics above, fine-grained cost metrics can be calculated for every step of the change management pipeline. For example, the costs of test service instances can be calculated by multiplying the duration of each step by the cost of the respective test instance. However, in most cases, the coarse-grained metrics provide enough information for optimization.

Coarse-grained cost metrics should be aggregated for the whole department and by service teams over a period of time, usually a month. If fine-grained cost metrics are used, they should be calculated per step and aggregated for the whole department and by service teams. The usual statistics should be calculated for fine-grained cost metrics, including the average, median, and standard deviation.

The total cost of the change management platform and all test instances over a period of time becomes a KPI for the entire department. The total costs aggregated by service teams over a period of time are the KPIs for the respective service teams. The fine-grained cost metrics calculated for each service and step should be used to optimize the process.

10.4. OPPORTUNITIES FOR MACHINE LEARNING

As soon as the required data is collected for all steps of the CI pipeline and change tracking is implemented efficiently, there will be plenty of opportunities to apply data analytics and machine learning. For example, inefficient steps may be found and optimized or low-risk changes can be predicted by sizes of commits and files modified by commits. The prediction of low-risk changes and high-quality committers may allow optimization of the change management process on a change-by-change basis.

11

FURTHER IMPROVEMENTS

There are certain improvements to the architecture and the process that we'd like to describe in this chapter. The improvements describe advanced techniques that not necessarily required for the first-time implementation of an efficient change management process. We would typically not recommend implementing these improvements right away, to let the organization fully adopt the cloud, microservices, and automation first. The improvements described below may also be difficult to implement, but they will provide additional benefits and can make the process even more lean and efficient, while retaining all necessary controls; ultimately, they can lead to true continuous delivery.

11.1. ROUTING FABRIC

The first advanced technique that can be implemented is a routing fabric. When we discussed blue-green upgrades of services deep in the service mesh in the section about service deployment, we mentioned that, if end-to-end testing is required and simple load balancing is used, the whole subgraph of downstream dependencies of a service should be upgraded.

In Figure 11.1., we assume that simple load balancing is used with the support of routing requests by HTTP request parameters. Even if the only service that needs to be upgraded is service B, if end-to-end testing, a canary release with a controlled customer group, or top-level A/B testing is required, a green instance of the web UI should also be created. There are two technical challenges in this case:

1. A load balancer without routing functionality cannot determine which instance of service B should receive a request.

Instead, a load balancer instance is created with the new service instance during the upgrade and the balancer sends requests only to the instances of the service of the same version. Routing logic is available and configured in this case only on the level of the global traffic manager.

2. Even if routing logic were available in the load balancer of service B, it wouldn't know about the end customer, so routing information will be absent in the request.

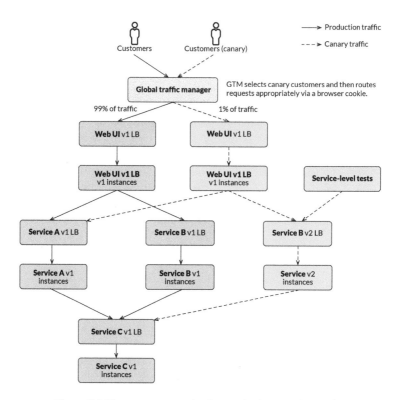

Figure 11.1. Blue-green upgrade of a service in a service mesh.

The obvious disadvantage of the approach above is that a blue-green upgrade of the most upstream service will require the recreation of the whole subgraph of services. For example, if service C should be upgraded, new instances of all services and the web UI should be created.

There are several major options to eliminate this disadvantage. First, the number of upgrades that require canary releases or end-to-end testing can be minimized or eliminated altogether. In this case, a new service deep in the graph can be upgraded with rolling or blue-green upgrades and the downstream dependencies will switch to the new version automatically. The problem with this approach is that sometimes end-to-end testing and canary releases are actually required.

The second approach is not to rely on deployments and upgrades as a method to do end-to-end testing or canary releases and instead rely on feature flags. In this case, the new instances can be upgraded, as described in the first approach, but the required features will be turned off. Feature flags can then be configured on the top-level service and passed as an HTTP request parameter to the upstream services. The challenge with this approach is the increased cost of creating and maintaining feature flags for all features.

Alternatively, a routing fabric can be implemented. The approach with the routing fabric is an extension of the second approach, in which each load balancer for each service has routing capabilities and services pass routing information in requests to the upstream dependencies.

A routing fabric requires implementation of several capabilities on the load balancing and application levels:

1. New instances of service load balancers are not created during blue-green upgrades. The same instance of the load balancer is reused.

2. All load balancers have routing capabilities and can route requests to specific groups of instances based on HTTP request parameters, like cookies or URL paths.

3. Applications need to pass certain HTTP request parameters to downstream dependencies for each request to make the routing information available throughout the services graph.

The exact method of passing routing information may depend on the capabilities of specific load balancers and applications. A naming convention for the routing fields should be established. In the example above, the name of the routing field was derived from the service name and the value was the version of the service. However, feature flag names and "on"/"off" values can be used instead.

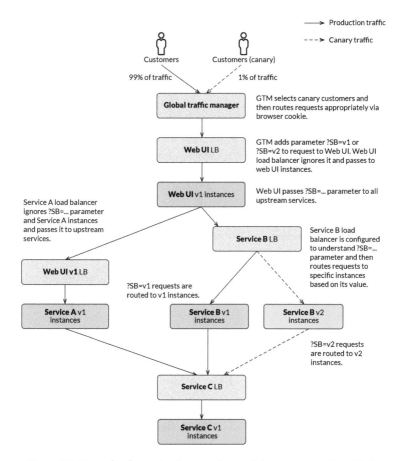

Figure 11.2. Upgrade of a service in a service mesh by using a routing fabric.

The routing fabric may also be used for end-to-end integration and user acceptance testing without the creation of additional test instances of components and services that are not being upgraded.

The selection of customers to participate in canary releases or A/B testing can still be implemented on the top level of the global traffic manager. Alternatively, a separate component integrated with the web UI can be used to select and manage such control groups.

The configuration of routing rules in load balancers can be done by a central load balancer if a separate custom routing and configuration dashboard is implemented. The configuration of routing in load balancers can also be merged with a traditional feature flags configuration to have consolidated control over what customers are using what functionality at any period of time.

The benefit of implementing the routing fabric is that it will allow end-to-end testing, canary releases, and A/B testing on the infrastructure level, not on the application level. It will minimize the usage of application-level feature flags and make application code development and maintenance easier.

At the time of writing, there are several technologies that enable routing fabrics for service meshes. Such tools include likerd, envoy, istio, nginx.

11.2. SINGLE ENVIRONMENT

The management of multiple environments for different services and different stages of the pipeline is problematic, leads to increased infrastructure costs and labor costs to support proper configuration of environments and instances within those environments, is more error prone because of potential misconfigurations, and is less efficient overall as a result of the longer CICD process. Minimization of the number of environments and instances typically leads to a more straightforward, transparent, and efficient change management process.

Ideally, from the deployment perspective, there should be only one environment that is able to serve all testing and production needs. Practically, the building of such an environment and segregation of traffic and access between the production and non-production instances may be difficult to implement. We will first describe the concept of a single environment and will then see how the concept can be applied in the real world to minimize the number of environments and service instances required for the change management process.

For the concept of the single environment, we will make two assumptions:

1. A service-oriented architecture is properly implemented: services have well-defined contracts, all service changes are independent, and services can be released to production in parallel.

2. A routing fabric is implemented to enable upgrades of services inside the service mesh.

3. Service virtualization is implemented for early testing of changes by developers.

We will start with the static state, when no changes are being requested and implemented (Figure 11.3.). At this point in time, there is only the production environment with production instances of services. These instances serve the production traffic from customers.

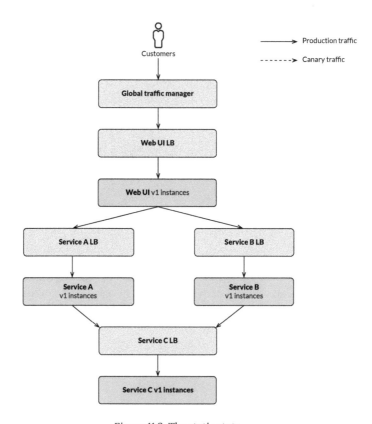

Figure 11.3. The static state.

The picture so far is trivial and describes a usual production environment. Now let's assume that two new requirements need to be implemented in service B and service C. The change management process works as usual until the new version 2 deployment packages are built for both services:

1. Requirements are created and approved in the requirements management tool.

2. The development teams implement new code for both services, execute the necessary tests locally by using service virtualization, and commit the code to the source code repositories. The code review passes and the commits are approved by the respective leads.

3. Deployment packages containing new versions of the services are built and are available for deployment.

The next step of the pipeline is service-level testing for the new versions of both services. Instead of the new versions being deployed to separate test environments, the services are deployed to production environments and connected with the production endpoints of upstream dependencies.

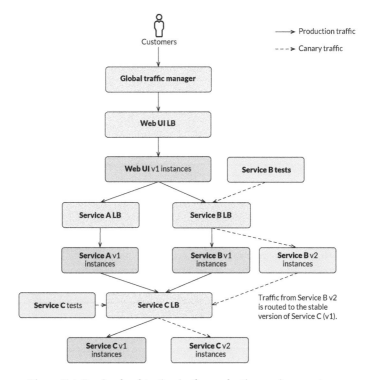

Figure 11.4. Service-level testing in the production environment.

Both services are tested independently. The new version of service B is actually connected with the stable version of service C (v1), because the service B team is not aware of the new version of service C and this version may not be good.

When the functional testing is done, profiling testing is initiated with the same instances of services.

Functional testing with production dependencies should not affect the functionality of production dependencies, with two assumptions:

1. Stress testing is not performed during the regular workload to avoid overloading production instances.

2. Services provide the necessary multi-tenancy capabilities to separate and protect the production data from the test data. For example, if service C manages sensitive production data, this data will not be available for the test instance of service B. Instead, service B should be able to generate test data that will not interfere with production or the other test data in service C. The segregation of data may be done by attributing specific data with the origin of the data and then implementing certain access controls to prohibit data being accessed by services with a different origin.

Profiling, stability, and even some level of performance testing may be performed as well. The stable components are working successfully in production, so they should be able to sustain the production load and stability issues by design, and they should have already been tested for compliance with non-functional requirements. The stable versions of services may also be overscaled, specifically to take into account the potential functional and performance testing in addition to regular customer traffic.

If a version of a service is rejected after failed testing, it may be destroyed and removed from the environment.

When the new version is ready to be released, it may start accepting canary traffic or regular production traffic. If both service B and service C are ready for release at the same, there may be four canary traffic configurations: both old versions, the new version of service B + old version of service C, the old version of service B + new version of service C, and both new versions. At some point, the new versions become stable versions and the traffic is switched away from the old versions completely.

Dependent changes in the service graph may also be implemented with a single environment, but this will require an understanding of which combinations of versions are good and which are bad. It should also be ensured that the test and canary traffic routing is configured appropriately. If there are multiple changes in the same service, multiple instances of a service may be deployed to the environment as well.

The additional benefit of having a single environment is that only a single service deployment package is required because the same instance of a service is being used for functional testing, performance testing, and production. This further minimizes the duration of the change management process.

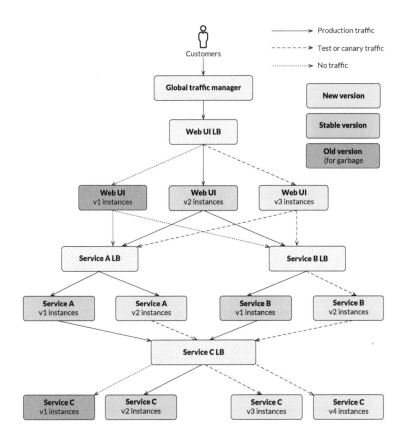

Figure 11.5. A single mesh of services in a single environment.

In the case of a single environment, all versions of services are joined in a single mesh (Figure 11.5.). This mesh then grows on one side as new versions are developed. On the other side, the mesh is reduced as old instances are no longer needed. One of the challenges with this approach is the identification of inactive instances that are no longer needed and no longer serve traffic. Although it may be hard to identify and remove such instances manually, it may be relatively easy to collect this information automatically on the basis of the configuration of the respective load balancers. Automatic garbage collection, similar to JVM or .NET garbage collection, can be implemented to identify, scale down, and remove unused instances of services.

The practical challenges related to the implementation of a single environment include:

1. Segregation of test and production traffic for compliance reasons.

2. Segregation of test and production data and implementation of multi-tenancy in services.

3. Segregation of access to test and production instances and infrastructure.

4. Implementation of service virtualization (mocks, stubs) for development testing.

5. Absence of robust tooling available on the market to manage the routing fabric, routing configuration, and garbage collection.

6. New and unusual process for most engineers working in an organization, especially an organization that is only just starting the transition to microservices, clouds, and continuous delivery.

All of these challenges may be overcome with custom implementation, but the cost of such implementation and the people risks need to be carefully evaluated. This is an advanced functionality, and it may be dangerous to use it as a starting point for implementing a continuous delivery process in an enterprise.

However, the challenges and risks of the single environment may be minimized significantly if production is separated from non-production. In this case, all functional and non-functional testing of all services may still be performed in a single environment with shared service instances. The production environment will still be separated into a separate infrastructure with separate access control and separate service instances. The implementation of such a single non-production environment with shared service instances may be a good step towards the optimization of infrastructure costs and the building of a more lightweight change management process.

11.3. ULTRALIGHT PROCESS

One of the goals of minimizing the number of environments or implementing a single environment is to optimize and shorten the change management process without sacrificing the compliance to base policies. As we discussed in the previous section, a single environment can significantly reduce the time that it takes to fully process and approve a single change. A reduction in this time will lead to fewer changes accumulating in the pipeline and a lower risk of deploying a change. Short change management cycles also mean that developers will switch context less often and will be more involved in following a change through all stages of its life cycle until the new version of a service that they implemented starts serving production traffic. This, in turn, will increase the motivation and productivity of the engineering teams, make the whole process more transparent, and allow the implementation of true Agile principles.

The implementation of a single environment is not the only method for achieving an ultralight change management process. Before we discuss other methods, let's review the key change management policies:

1. A change can be requested only by authorized personnel.

2. A change should be approved by the lead of the team that prepares and develops the change.

3. A developed change should be approved by the quality assurance lead.

4. A developed change should be approved by the performance testing lead.

5. A developed change should be approved by the requestor.

6. A developed change should be approved by the security lead.

7. A developed change should be approved by the operations lead.

8. A developed change can be implemented in production only after all approvals have been collected.

9. An implemented change should be reviewed by the requestor, operations, and the quality assurance lead.

10. An audit log should be available for all changes end-to-end.

The most time consuming processes are typically related to the policies for the collection of approvals from various leads, because these approvals require the execution and analysis of tests. The key in the implementation of an ultralight process is to follow all policies but to understand what they really mean and adjust the process and tooling.

Let's review the process in Figure 11.6. as an example.

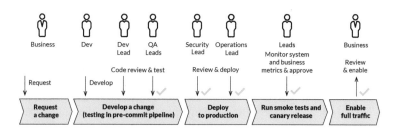

Figure 11.6. An ultralight change management process.

The initial creation and approval of the business requirement and the development of the change are performed as usual.

However, the later pipeline is different in the following ways:

1. Functional tests and security validations are executed during the pre-commit process. The tests should be fast, so only smoke tests may be selected for execution.

2. Both the development and QA leads should approve the commit, which is then committed to the master branch.

3. The build and packaging step is fast, so it is executed as usual.

4. The whole testing pipeline is skipped. Instead, the new version of the service is deployed to production in blue-green mode.

5. After deployment to production, basic smoke tests are executed and a canary release is performed with a small percentage of customer traffic routed to the new version.

6. System and business metrics are monitored during the canary release. If something goes wrong, the traffic is immediately switched back to the stable version.

7. If the canary release is successful, the business users gradually switch more traffic to the new version until the new change is fully enabled and old version is destroyed.

The process still satisfies all of the change management policies above, but it can be executed extremely fast. It is important to note that the actual deployment of a new version of a service to production doesn't correspond to the "production deployment" statement in the policy. Instead, the enabling of full traffic is the actual "production deployment." The testing policies in this case are only partially implemented with automated tests. Automated tests are used only to ensure that the change doesn't have critical and obvious defects. The real end-to-end testing is performed with the canary traffic. If system and business metrics are selected correctly, the negative impact to customers in the case of defects is minimal, because the change will be rejected and the traffic will switch back to the stable version.

Such an ultralight process may be extremely efficient, but it has a number of challenges. First, there is still the possibility that some customers may see defects, even temporarily. The second challenge is that this process requires mature team members who understand their responsibilities and the cost of defects. Therefore, this process may not be a good fit for truly mission-critical services and cases where the service team is not ready for it. However, even these challenges may be mitigated if the process is implemented for select services and teams or if the process is implemented as a "happy path" process. A "happy path" process works for changes by default and, if a defect is found during the canary release, all future changes will go through a more traditional and heavyweight process until the defect is fixed. The choice of whether to use a "happy path" process or traditional process may also be decided on the basis of the developer and the size of the change. For example, if a developer is known to provide very high quality changes and the change touches a small number of lines of code in non-critical source code files, the change may go through the ultralight process.

There may be other opportunities to implement a lightweight change management process. These opportunities for improvement come from technology innovations and new tools being available on the market, as well as the maturity of service development teams. The key to building a good process is to always reevaluate it and take a fresh look at the change management policies and the reasons why they exist.

———————————————

12

CONCLUSION

In this book, we have described a blueprint for the implementation of continuous delivery. We focused on enterprise IT environments, which are starting their journey from traditional software change management processes, monolithic applications that are deployed in private data centers, and mostly manual processes. We discussed the reasons why such transformation is necessary and why now is the time to do it. We approached the transformation from the perspective of retaining core policies, while rethinking and replacing traditional processes with new ones that are enabled by clouds, microservices, ubiquitous automation, and the DevOps culture. We believe that such an approach will lower the barrier for transformation and justify many improvements that may otherwise seem too risky and dangerous to the existing teams. During the book, we have also tried to include actual detailed implementation patterns, techniques, and technologies to give actionable recommendations in addition to general conceptual considerations and advice. In the end, we would like to leave the reader with several key messages.

First, no matter how safe the approach to transformation is, change is always hard. The movement toward a true service-oriented architecture and service-oriented organization may be difficult, but in our opinion, it is absolutely necessary to gain efficiency in the long term. It is especially important for large companies, because it allows them to scale up more easily or, at least, manage organizations and systems at scale. The implementation of cloud infrastructure, microservices architecture, and automation in all aspects of software lifecycle management activities helps with the migration toward a service-oriented organization. A review and adjustment of the roles and responsibilities of different organizations and an understanding of the interfaces and contracts between them are key factors for success.

Second, when you are implementing continuous delivery, focus on careful analysis of existing change management policies, but do not get stuck with existing processes and procedures. In most cases, existing policies stay valid no matter how fast you would like to deliver changes to production. However, there is a thin line between an existing policy and a legacy process, and the two of them are confused too often. Whereas policies should remain, legacy processes should not become blockers for transformation. We strongly recommend investing time and understanding the motivation behind the policies and then designing a new process from scratch that will satisfy the policies. Sometimes, if necessary, even low-level policies may be changed if new ones would satisfy higher order policies and controls.

Finally, optimization and improvement is an ongoing process. It is important to start tracking metrics and define the KPIs of the process so that it can be optimized and improved. Some improvements will be straightforward; others will be more radical and may require more breaking changes. We discussed some of the radical improvements that can bring significant efficiency at the end of the book, but new ideas and improvements will emerge with a better adoption of continuous delivery in the industry. At the time of writing, the implementation of continuous delivery and DevOps is still an art, and we hope it will become more scientific as more data is collected from the industry and in each company.

REFERENCES

[1] D. Edwards, "What is DevOps?," February 23, 2010, http://dev2ops.org/2010/02/what-is-devops/.

[2] C. Larman and B. Vodde, Scaling Lean & Agile Development: Thinking and Organizational Tools for Large-Scale Scrum, Pearson Education, Boston, MA, 2009.

[3] R. Bias, "The History of Pets vs. Cattle ... & Using it Properly," September 30, 2016, https://www.slideshare.net/randybias/the-history-of-pets-vs-cattle-and-using-it-properly.

[4] "Cloud Computing," Wikipedia, last edited May 25, 2018, https://en.wikipedia.org/wiki/Cloud_computing.

[5] J. Lewis and M. Fowler, "Microservices: A definition of this new architectural term," March 25, 2014, https://martinfowler.com/articles/microservices.html.

[6] P. Hammant, "An Ontology: Component vs Class vs Object vs Service vs Application vs Process vs Library, etc.," January 31, 2018, https://paulhammant.com/2018/01/31/ontology-component-vs-class-vs-object-vs-service-vs-application-vs-process-vs-library-etc/.

[7] "DevOps," Wikipedia, last edited May 4, 2018, https://en.wikipedia.org/wiki/DevOps.

[8] "Continuous Delivery," Wikipedia, last edited May 20, 2018, https://en.wikipedia.org/wiki/Continuous_delivery.

[9] M. Fowler, "Continuous Delivery," May 30, 2013, https://martinfowler.com/bliki/ContinuousDelivery.html.

[10] "Change Management (engineering)," Wikipedia, last edited March 27, 2018, https://en.wikipedia.org/wiki/Change_management_(engineering).

[11] "Change Management (ITSM)," Wikipedia, last edited December 16, 2017, https://en.wikipedia.org/wiki/Change_management_(ITSM).

[12] "Site Reliability Engineering," Wikipedia, last edited June 13, 2018,   https://en.wikipedia.org/wiki/Site_Reliability_Engineering.

[13] "Change Control Board," Wikipedia, last edited December 8, 2016, https://en.wikipedia.org/wiki/Change_control_board.

[14] "The Great Microservices Vs. Monolithic Apps Twitter Melee," High Scalability, July 28, 2014, http://highscalability.com/blog/2014/7/28/the-great-microservices-vs-monolithic-apps-twitter-melee.html.

[15] "AWS Identity and Access Management (IAM)," AWS, accessed June 13, 2018, https://aws.amazon.com/iam/.

[16] "Cloud Identity and Access Management," Google Cloud, accessed June 13, 2018, https://cloud.google.com/iam/.

[17] "Azure Active Directory," Microsoft Azure, accessed June 13, 2018, https://azure.microsoft.com/en-us/services/active-directory/.

[18] "Creating and Managing Projects," Google Cloud, accessed June 13, 2018, https://cloud.google.com/resource-manager/docs/creating-managing-projects.

[19] "Account," AWS, accessed June 13, 2018, https://aws.amazon.com/account/.

[20] "Using Resource Hierarchy for Access Control," Google Cloud, accessed June 13, 2018, https://cloud.google.com/iam/docs/resource-hierarchy-access-control.

[21] "IAM Best Practices," AWS, accessed June 13, 2018, https://docs.aws.amazon.com/IAM/latest/UserGuide/best-practices.html.

[22] "Using Cost Allocation Tags," AWS, accessed June 13, 2018, https://docs.aws.amazon.com/awsaccountbilling/latest/aboutv2/cost-alloc-tags.html.

[23] "Creating and Managing Labels," Google Cloud, accessed June 13, 2018, https://cloud.google.com/resource-manager/docs/creating-managing-labels.

[24] "Managing Instant Access Using OS Login," Google Cloud, accessed June 13, 2018, https://cloud.google.com/compute/docs/instances/managing-instance-access.

[25] R. Rowan, "The best article I've ever read about architecture and the management of IT," October 12, 2011, https://plus.google.com/+RipRowan/posts/eVeouesvaVX.

[26] "Conway's Law," Wikipedia, last edited May 6, 2018, https://en.wikipedia.org/wiki/Conway%27s_law.

[27] "Inverse Conway Maneuver," ThoughtWorks Technology Radar, accessed June 13, 2018, https://www.thoughtworks.com/radar/techniques/inverse-conway-maneuver.

[28] "Separation of Duties," Wikipedia, last edited April 10, 2018, https://en.wikipedia.org/wiki/Separation_of_duties.

[29] V. Driessen, "A successful Git branching model," January 5, 2010, https://nvie.com/posts/a-successful-git-branching-model/.

[30] S. Chacon, "GitHub Flow," August 31, 2011, http://scottchacon.com/2011/08/31/github-flow.html.

[31] P. Hodgson, "Feature Toggles (aka Feature Flags)," October 9, 2017, https://martinfowler.com/articles/feature-toggles.html.

[32] T. Preston-Werner, "Semantic Versioning 2.0.0.," accessed June 13, 2018, https://semver.org/.

[33] C. Richardson, "Pattern: Service Registry," accessed June 13, 2018, http://microservices.io/patterns/service-registry.html.

[34] "Rolling Release," Wikipedia, last edited August 26, 2017, https://en.wikipedia.org/wiki/Rolling_release.

[35] M. Fowler, "Blue-Green Deployment," March 1, 2010, https://martinfowler.com/bliki/BlueGreenDeployment.html.

[36] "StatefulSet Basics," Kubernetes Tutorials, accessed June 13, 2018, https://kubernetes.io/docs/tutorials/stateful-application/basic-stateful-set/.

[37] D. Sato, "Canary Release," June 25, 2014, https://martinfowler.com/bliki/CanaryRelease.html.

[38] W. Morgan, "What's a service mesh? And why do I need one?," April 25, 2017, https://blog.buoyant.io/2017/04/25/whats-a-service-mesh-and-why-do-i-need-one/.

[39] M. Fowler, "Mocks aren't Stubs," January 2, 2007, https://martinfowler.com/articles/mocksArentStubs.html.

[40] M. Fowler, "Test Double," January 17, 2006, https://martinfowler.com/bliki/TestDouble.html.

[41] C. Rosenthal, L. Hochstein, A. Blohowiak, N. Jones, and A. Basiri, *Chaos Engineering: Building Confidence in System Behavior through Experiments*, O'Reilly Media, Sebastopol, CA, 2017; available as an ebook from https://www.oreilly.com/webops-perf/free/chaos-engineering.csp.

CPSIA information can be obtained
at www.ICGtesting.com
Printed in the USA
LVHW072220220221
679662LV00037B/1425